Given their capabilities, it's no surprise that computer terminals are replacing typewriters and cash registers in every type of business today. They're being used for letter writing, billing, order entry, and credit checking, as well as for filing and retrieving information and a host of other activities. But for the manager not familiar with these new technologies, the task of selecting and using a computer system seems an almost impossible one. In this book, Harry Katzan, Jr., one of the most respected authorities in the computer field, provides you with the information you will need to plan, design, and implement a system —one that not only meets your present needs but can be expanded to handle your future requirements.

The book is divided into three sections. In the first—an **Introduction**—Katzan discusses the revolution in electronics and how it affects organizations. He also examines the different types of systems available and the functions each is capable of performing. In the second section—**Basic Concepts**—he explains in layman's terms the fundamentals of computers, software, and communications as they relate to the use of computers. The third section—**Topics in Office Automation**—provides detailed coverage of word processing, electronic mail, systems concepts, and management strategy. In addition, the book includes an extensive bibliography and an up-to-date glossary containing over 900 computer terms.

Ultimately, the real advantage of implementing office automation is the resulting increase in office productivity. This book will not only provide you with an understanding of the field, but will also enable you to recognize recent advances in technology and apply those advances in your own office or company.

Office Automation

Office Automation

A Manager's Guide

Harry Katzan, Jr.

amacom

AMERICAN MANAGEMENT ASSOCIATIONS

Library of Congress Cataloging in Publication Data

Katzan, Harry.
  Office automation.

  Bibliography: p.
  Includes index.
  1. Office practice—Automation.   I. Title.
HF5548.2.K385 1982   651.8'4   82-71309
ISBN 0-8144-5752-5

First Printing

Preface

The term "office automation" is in the news these days as a means of increasing total office productivity. The issues involved are complicated by the fact that the terminology means different things to different people—technologists as well as executives.

No matter how the terms are defined, office productivity has three dimensions: technical, behavioral, and managerial. It is impossible to implement an effective office automation program without education, training, and planning in all three.

This book shows how to utilize technology to improve office functions. By using the information presented here, you should be able to recognize recent advances in relevant technology and to apply these advances to the office environment. This book should be useful to executives, administrators, and managers who are interested in improving productivity through office automation.

Harry Katzan, Jr.

Contents

PART I

Introduction

CHAPTER 1

The System Is the Solution

The term "office automation" is in the news these days as a means of increasing total office productivity. In fact, computers are even bigger news: personal computers, microcomputers, microprocessors, word processing, electronic mail, executive work stations, and on and on—the new terminology is overwhelming, and the underlying technology is practically beyond comprehension. Office automation is inevitably related very closely to computer technology, as both represent recent advances in microelectronic technology. These advances have surfaced at a time when worldwide unemployment and economic growth are of concern. Thus the technology looms as a double-edged sword, since the threat of additional unemployment represent significant side effects in most industrial countries.

The Micro Revolution

Some people have called it the "second industrial revolution," with an impact potentially greater than that of the original industrial revolution that moved people from the farms to the cities and workers from the fields to the factories. Characterized by computer-controlled machines ranging from intelligent typewriters to electromechanical robots, this revolution represents a multibillion-dollar industry that has already brought sweeping social changes in how we live, how we work, and how we enjoy our leisure hours.

In the short span of only a few years, we have all seen or read about automated factories, computerized newspaper composition, pocket calculators, hand-held computers, electronic games, digital watches, language translators, and automobile electronics—and this is only the tip of the iceberg. But probably more significant than all of those is the simple fact that computers have now invaded the office, and the traditional distinction between data processing and office machinery no longer exists.

Return on Information

Most people associate data processing with billing, record keeping, and other diverse forms of accounting. In spite of a full complement of horror stories ranging from billing errors to computer crimes, the impact of computers has always been remote, because the normal relationship to the computer was always through an intermediary—a representative of the data processing group or a specially trained user person. Problems, if they existed, clearly involved getting the correct information into the computer. Somehow the output always took care of itself, even though understanding it could sometimes prove to be a challenge.

With the automated office, the situation is reversed, since it effectively involves getting information "out of the machine" in a timely and efficient manner. Because computer intelligence can be built into typewriters, copiers, communications devices, and other machines that perform office processes, we are at present looking for a "return on information." We have an investment in information, and the challenge of the 1980s is to use it well.

Small Is Beautiful

The key to work effectiveness and office productivity is people—but as we have seen in the 1970s, good management practices are clearly not sufficient. In the last decade, factory worker productivity increased 80 percent, while the productivity of the white-

collar worker increased only 4 percent. Reasons for the dispropor-
tionate rate of increase vary among organizations; however, one
point remains unmistakably clear: *Any increase in worker productivity
is directly related to the average capital investment per workplace.* In the
last decade, businesses invested $25,000 to $40,000 for each factory
worker, while a much more modest amount, perhaps as low as
one-fourth of that amount, was invested in the office worker. The
implications are obvious. But office automation is not as simple as
installing new equipment.

In discussing what we need from the scientists and technologists, the noted economist E. F. Schumacher writes in *Small Is
Beautiful* that we need methods and equipment which are (1) inexpensive enough to make them individually accessible, (2) designed
for small-scale applications, and (3) compatible with our need for
creativity.

Clearly, the microelectronics industry has satisfied all these
needs through integrated circuits and computers on a chip, while
computer scientists have enabled this computer power to be applied to all aspects of modern society. Yet the computer must be
integrated into the very inner workings of the organization—that is,
the office—to realize benefits in productivity and efficiency. Just
how this is done determines the effectiveness of the automated office. In short, it is not the technology that is the solution, but rather
the system itself.

The Environment for the Modern Office

The modern office is conveniently referred to in the popular lit-
erature as the "office of the future," or the "automated office." Neither of these terms is even remotely accurate—the so-called office
of the future exists now, and it is becoming increasingly clear that
most office work will never be truly automated. Repetitive tasks
will be circumvented, and detailed work that is error-prone will be
eliminated. Judgmental work, at least for the foreseeable future,
will be performed without automation. But the key benefit will be
that office people "work smarter," something that will ultimately

affect all of us. The nature of office work will most certainly change, but the basic tasks will not be eliminated.

In the present context, both a name and a definition are needed for the subject at hand, as it is commonly associated with "data entry" devices and with "word processing" machines. Obviously, typewriters and dictating machines ultimately involve word processing and data entry operations; however, something more esoteric is normally expected when the modern office is involved. Recognizing its obvious limitations, the subject will be apologetically referred to as "office automation" whenever a convenient label is required.

The term "office automation" refers to the use of a computer in an office environment to facilitate normal operating procedures. The impact of office automation upon work flow can be very small or very great, depending on the extent to which organizational structures are affected.

Some of the fundamental issues that are commonly raised with regard to office automation strongly imply that the use of advanced technology does not translate directly into increased office productivity. Moreover, even getting started with an office automation program may be a major obstacle, since routine changes in normal office functions almost always have to be justified. One is reminded here of Parkinson's Law of Triviality, which says that the time spent on any item of an agenda will be in inverse proportion to the sum involved. Thus an organizational unit initiating an office automation pilot project perhaps with a stand-alone word processor may expect to encounter rebuttals such as these:

- The increased productivity associated with office automation will increase the proliferation of information—that is, more paper in a system already saturated with loaded file cabinets.
- The long-term result of office automation will invariably be an increase in the number of people doing the same amount of work.

Clearly, either disaster can easily result if office routines are not adjusted accordingly and if proper familiarization and training programs are not implemented. However, the primary objective of office automation is to increase the bottom line, something that should rightly be taken into consideration where a total office automation project is undertaken.

Another important consideration is that doing what is being done now—producing letters and reports and doing office communications and filing and retrieving information—in a more productive manner may not be desirable from an organizational point of view. Thus the overall gain from an automated office is that less of such work is performed, and the major benefit lies in the reorganization and restructuring of office functions. Perhaps the most significant aspect of the office environment is that office work is not as well structured as accounting, scientific, and engineering work, where computer technology is used regularly.

Office automation may be difficult to implement because of the diversity of office functions, and this fact has led many people to believe that person–machine interaction is the prime ingredient in a successful system. While person–machine interaction is definitely important, a focus on the whole system is needed to ensure that the people involved can have comfortable experiences and will want to use office automation equipment.

Functions Affected by Office Automation

Technology is introduced into an office environment to enhance the functions performed by office workers, taken here to include anyone who works in an office from executives to clerk trainees. Initially, most office automation systems involve automated functions that are logically unchanged from their manual counterparts. The replacement of a traditional office typewriter with a stand-alone word processor is a case in point. Subsequent stages of growth in office automation call for an integration of the various functionalities. (*Functionality* is a term commonly used in the computer field, comprising the functions performed, how these functions are used, and their results.)

The functions normally associated with office automation incorporate the following benefits:

- An increase in the speed of conventional office functions.
- An improvement in the quality of conventional office functions.
- A decrease in the level of manual operations and other physical activity.
- An increase in cooperative management behavior.

The notion of speed is commonly associated with automation and productivity both in and out of the office. In the office, automation increases the speed of preparing letters and reports, of long-distance communications, and of everyday office communications. The time needed to go from a draft copy to a final document is minimal with office automation in comparison with traditional means. Electronic communications systems, which include message systems, digital voice distribution, digital facsimile, and other forms of conferencing, facilitate long-distance communications and provide enhanced capabilities for storage and distribution. Data entry can be improved through optical character recognition equipment that permits existing documents to be entered directly into an office information system without rekeying. Information retrieval and electronic filing systems provide almost instantaneous access to large amounts of information, and automated filing systems can decrease the time ordinarily spent on manual office operations while providing a high level of operational consistency at the same time.

The high level of quality inherent in office automation is accepted without question. Perfect copies produced by word processing, and an increase in completed communications available through electronic message systems, are evidence of the qualitative benefits to be gained through office automation.

The term "automation" strongly connotes a decrease in manual operations and physical activities that range from correcting, filing, and copying to ordinary commuting and travel. Telephone and travel time can be reduced with message and conferencing systems, respectively. Information can be retrieved by index entries or context-dependent keys without performing manual lookup operations.

The success of an organization ultimately depends on the cooperation of people, and office automation can provide more cooperation between managers by decreasing the time spent on routine tasks and by permitting high-priority activities to be performed with great efficiency. Clearly, benefits obtained through word processing, electronic filing, and electronic communication facilities can free managers to concentrate on more important aspects of organizational work. And decision support systems together with various forms of teleconferencing allow management to collaborate to create dynamic organizational rapport where it was not previously possible.

Electronic communication systems demonstrate very well how cooperative behavior can be improved. A common time waster is "telephone tag": manager A calls manager B; manager B is tied up, so a message is left; manager B later returns the call, but now manager A is busy; and so forth. With an electronic message system, routine communications can be performed through the office computer. Does this mean that the telephone will be unnecessary? Clearly not. But it does mean that the telephone will be used more efficiently when an immediate response is needed. Parenthetically, electronic message systems are particularly appropriate for routine communication between parties in different time zones. Between New York and San Francisco, for example, a time window of approximately four hours is all that is available for telephone communications under ordinary circumstances. Between the United States and Europe, the time window is even shorter.

Knowledge and Support Workers

Office automation specialists distinguish between knowledge workers and support workers. Executives, managers, professionals, and information administrators are classified as knowledge workers because they use a great amount of information during their everyday activity. Support workers enter, process, and otherwise manipulate the information used by the knowledge worker. A company president, for example, is a knowledge worker. A "typing pool" clerical person would be a support worker.

Office automation is an end-user discipline. Seven basic categories of system users have been identified:

- Managers of managers.
- Managers of professionals.
- Managers of clericals.
- Professionals.
- Administrative clericals (for example, executive secretary).
- Production clericals (for example, order-processing worker).
- Typing-pool clericals.

Managers represent only three of the seven categories, but their impact on the office environment vis-à-vis the benefits of office automation is generally felt to be much greater than the impact of the other three categories.

Productivity

The productivity of people in an office environment depends on a continuous interplay among three major activities: learning, thinking about how to improve, and adaptation to new processes and procedures. Management is responsible for providing the necessary resources, and management is correspondingly accountable for the result. Within the domain of responsibility and accountability, however, both knowledge and support workers must combine learning, thinking, and adaption to realize productivity benefits.

Learning means recognizing what a worker's job really is and understanding the potentialities of the available resources. *Thinking* should involve analyzing the tasks with the productive use of the resources in mind. *Adaption* consists of changing the work environment to realize the improvements generated in stages one and two, learning and thinking.

Modern managers recognize that the management of productivity involves four key areas: capital, critical physical resources, time, and knowledge. As a separate area, the management of knowledge involves the total organization, because information is intertwined throughout the diverse network of people and the internal structure. Moreover, knowledge productivity cannot be achieved simply by replacing people with information-processing machines. Thus a capital investment program by itself is not sufficient; the effective management of people will remain a key ingredient.

The Two-Headed Monster

Society tends to view some professional employees as two-headed monsters: people you can't live with and can't live without. People in this category are characteristically more dedicated to their discipline than to the welfare of the organization. Another manifestation of two-headedness is resistance to be held accountable for results.

The point is that systems and concepts can be as two-headed as people. And office automation can easily be such a two-headed monster, unless key people are held accountable for results, both qualitatively and quantitatively, and at a reasonably good cost/benefits level. Once an organization "cuts over" to office automation, it cannot easily turn back the clock. The proliferation of paper and people may be the end result if needed training, management support, and systems integration are not provided.

The Waves of Computers

Office automation represents the second wave of computers in modern organizations (not to be confused with A. Toffler's second wave of civilization). Both the first and the second wave profoundly changed the office environment.

The first wave was data processing. With data processing, the major emphasis is on the entry, processing, and output of data. A primary objective in data processing is to keep the computer busy, while at the same time providing the wherewithal for managing large amounts of information.

With office automation, the major emphasis is on using computers to help people. A primary objective in office automation is to allow people to work efficiently while providing access to information in the office environment.

The distinction between data processing and office automation is not as well defined as most people think. In recent years, there has been a move toward integrated systems that permit access to large amounts of information from the office and the supporting of traditional office functions by computers previously restricted to the data processing domain.

Summary

Office automation is being offered as a means of increasing the productivity of the modern office and allowing the total office system to operate at a higher level of efficiency. The move to office automation largely results from the microcomputer revolution that has entered other areas of our lives.

While office automation is closely akin to data processing, it represents a different modality with an emphasis on getting information out of the computer. It improves the return on information generated and stored with traditional data processing.

A key aspect of office automation is that it is user-oriented: methods and equipment must be designed not only to be inexpensive and productive but also to satisfy people's need for rewarding human experiences.

Technology is introduced into an office environment to enhance the functions performed by office workers. The notion of an "office worker" is extended in the sense of office automation to include the knowledge worker and the support worker. In varying degrees, both classes of worker can expect benefits in the following areas:

- Increases in the speed of traditional office functions.
- Improvement in the quality of office work.
- Decreases in the level of manual operations and physical activity.
- Increases in cooperative management behavior.

In order to be truly effective, office automation should benefit the organization as a whole, as reflected in the "bottom line." Thus integrated systems are the main ingredient to successful office automation. "The system is the solution" in the office of the future; however, local productivity improvements throughout word processing and electronic messaging can be achieved along the way.

SUGGESTED READING

Dertouzos, M. L., and Moses, J., *The Computer Age: A Twenty-Year View*, Cambridge, MA: MIT Press, 1979.

Drucker, P. F., *Managing in Turbulent Times*, London: Pan Books, 1980.

Evans, C., *The Micro Millennium*, New York: Viking, 1979.

Parkinson, C. N., *Parkinson's Law and Other Studies in Administration*, Boston: Houghton Mifflin, 1957.

Toffler, A., *The Third Wave*, New York: Morrow, 1980.

CHAPTER 2

The Dimensions of Office Automation

One way of looking at office automation is that it is the process of using people, procedures, and equipment within an office environment. While an organizational framework is implicit in most informal definitions of office automation, it is not specifically required for a functional viewpoint. A researcher working alone in a remote location or a person completely independent of organizational ties, such as an author, can effectively use many of the facilities ordinarily classed within the domain of office automation. Because office automation is not specifically a product but rather a set of procedures together with appropriate equipment, it is useful to categorize the functions performed in an office environment in order to identify the classes of products that could be useful in the various cases.

Components of an Automated Office

The functional components in an automated office can be placed in five classes:

Entry/generation system.
Transport system.
Storage system.

Retrieval/query system.
Output/distribution system.

The classes are *not* mutually exclusive and are intended to be suggestive of the various office functions that are performed.

The *entry/generation system* is a means of getting information into the pipeline. In a data processing environment, the functional area would represent a data entry type of operation. In an office automation environment, the entry/generation class of operations would include word processing, optical character recognition (OCR), facsimile technology, computer input microfilm, digitized voice, video, graphic input, and possibly even an "intelligent" copier of some type.

The *transport system* is used to move information from one location to another. In an office environment, this function also entails the distribution and dissemination of information. In general, this class of operations would include electronic mail, data networks, and various forms of video conferencing and telecommuting. Electronic mail tends to be a broad category and includes message systems, facsimile, voice distribution, and other forms of communication and associated technology.

The *storage system* involves the storage of letters, reports, documents, messages, and schedule/agenda information, as well as facilities inherent in specialized devices such as computer output microfilm (COM), digitized voice, and graphics. Newer technologies such as video disk and electronic filing also fall in this category.

The *retrieval/query system* is used to access the storage system through end-user devices and software facilities. Multiple-use terminals and specialized query languages highlight the operations in this category. Methods of accessing voice, video, and graphic information are also relevant to retrieval/query systems. The notion of an integrated system is particularly relevant to the functionality inherent in query and retrieval. Through the integration of office automation and data processing facilities, a reference can be made to information stored in a data base for purposes of incorporating that information into office automation documents.

The *output/distribution system* is used to generate documents in a "noncomputer" form. Printing, duplicating, folding, binding, photocomposition, and so forth are activities belonging to this system. The term "reprographics" is commonly used for output activities that precede the distribution function, which may be incorporated into the storage or transport system. In many cases, a document is transported, stored, and then distributed in a remote environment. An advanced office concept presently being considered in some large organizations involves the idea of a "printing and distribution" center, located in a separate facility for the express purpose of handling *all* output functions ranging from copying and duplicating to addressing, packing, and sending.

Grouping the functional components in an automated office in this way is not intended to be definitive as in an academic treatment of the subject matter. The various other functions are clearly inherent in each aspect of office technology in the sense that a particular technology, such as electronic mail, is used to enhance one or more specific functions, such as office communication in this particular case.

Office Functions

It is useful to view office work as a set of processes and a set of office functions. The processes can be those included in traditional management work or something performed by a professional, administrative, or clerical person. Office functions are activities *usually* performed as part of the various processes. The word "usually" should be taken seriously in this instance, since no simple tabulation of functions can completely describe the complex range of behavior contained in everyday situations, in the same way that no office automation system will be able to automatically solve every office problem. However, the effective use of office automation can provide the time and resources necessary for knowledge and support workers to handle problems that are not taken care of in an automated manner.

A representative set of processes performed by managers, professionals, and clericals includes planning, organizing, and controlling. Auxiliary processes, such as staffing, coordinating, monitoring, policy formulation, and decision making, can be regarded as subordinate, even though an alternative classification could easily interpret them as major functions. To a large extent, the three main processes also apply to clerical workers in a modern office environment, since the planning, organizing, and controlling of one's own work has become commonplace in the modern office.

The office functions that exist within these processes are:

Generating information.
Modifying information.
Gathering information.
Filing information.
Retrieving information.
Analyzing information.
Communicating information.
Output and distribution of information.

The fact that each activity involves information is obviously very significant and lends credence to the assertion that office automation is centered on the knowledge worker and the support worker. Now, within each of these functional areas, there exists a technology as well as a set of relevant procedures. The technology provides the mechanism for doing the respective function, but by itself it is not sufficient. The procedures represent the manner in which the technology is used, and can be divided into two categories: (1) procedures that for the most part apply to a function, such as word processing, but do not vary very much between different organizations, and (2) procedures that are specific to a given organization and represent a "management style" or a "way of doing business."

Even within the same organization, it is likely that two people will use the same facility in a different manner. For this reason, it is important that people collaborating on a project agree on procedures to be used for that project (such as methods of filing and accessing information), even if they use different procedures when working alone.

Word Processing

The term *word processing* refers to text preparation through the use of a computer or its equivalent. Historically, it referred to any process that involved textual information, such as copiers, typesetters, and automatic typewriters. As a computer concept, which is the current context, word processing refers to a text editing system in which a document, such as a letter or report, is typed and recorded on a magnetic recording medium.

The concept of word processing originated with an electric typewriter connected to a magnetic tape for storage. The tape permitted text to be saved for correcting, revision, and reprinting. Corrections and revisions could be made without retyping an entire page, and multiple error-free copies could be generated, each giving the appearance of an individually typed letter, report, or document. Subsequently, methods were developed for inserting names, addresses, dollar amounts, and other information in the body of the text so that word processing could be used to generate form letters. In the ensuing years, the recording media for word processing have evolved through magnetic tape, magnetic cards, tape cassettes, diskettes, and hard disks.

Modern word processing systems permit textual information to be entered, edited, and reviewed through a CRT/keyboard device, and small flexible diskettes or larger rigid disks can be used for permanent storage. A micro computer or minicomputer is used for processing, and the editing functions permit text to be rearranged, inserted, deleted, or replaced.

The key element in the word processing cycle is the output functions, which may include printing, communications, or some form of photocomposition. The output function is covered in Chapter 3 together with the delineation of other word processing functions.

The hardware components that make up a modern word processing unit are a keyboard, a printer, a display screen, a storage device, and a computer (known in word processing terminology as a processor). This collection of equipment is known as a *workstation,* which may exist singly as a stand-alone unit or may reside in one of the following configurations:

- A set of workstations electrically connected to share a printer. In this case, each distinct workstation has a processor, a keyboard, a display screen, and a storage device, but only one unit physically has a printer, which is shared among the other workstations in the system.
- A set of units containing only a keyboard and display screen connected to a centralized computer resource. This is known as a "shared logic" system, in which a minicomputer controls the storage, editing, and printing functions.
- A stand-alone word processing unit or a cluster of units connected to a data processing computer. This is generally classed as an "integrated system," because several office automation functions (such as word processing and electronic mail) are commonly available through the use of a single "universal" workstation.

A document prepared with a word processing workstation can be transmitted to another location in one of three ways:

1. By physically removing the storage medium (the diskette in most cases) from one system, physically carrying it to the other system, and inserting it in the corresponding storage device.
2. By transferring the document electronically from one location to another and storing it on the viewing end.
3. By accessing the document electronically in a centralized location, as would be the case with word processors connected to a centralized computer or to a shared logic system.

It is obvious that routine procedures must be established in order for people to work collaboratively on a document; however, the main advantage to this type of operation is that the time needed to perform traditional office work is diminished.

Optical Character Recognition

Optical character recognition (OCR) refers to the physical process of reading documents prepared on conventional office typewriters for entry to and storage in a computer system. While specific typing conventions and character sets are necessary for valid entry of information in this manner, the technique provides a means of "direct input" to the computer and of reducing the backlog of documents that normally exists during office system startup. In short, the use of OCR equipment saves time and keystrokes.

The problem addressed by optical character recognition is as follows. As an organization develops an operational office system through word processing and document storage facilities, either locally in the workstation or in a remote computer system, it becomes obvious after only a short period of time that a lot of the information needed exists only on paper. If the paper system were allowed to exist and all new documents were entered into the office system, it would take some time before a working set of documents was established to eliminate the need for running two large systems in parallel. Even then, the question would arise about documents produced in nonautomated offices. In short, this is the backlog problem.

Through the use of optical character recognition equipment, a stack of documents could be read into the office system and stored magnetically in a storage medium, probably disk. The normal office system functions could then be applied to the OCR-generated information as though it were entered into the system manually.

Another possibility, of course, is to continue producing and gathering typewriter-generated documents in the future, as is commonly the case at the present time in most offices. These documents could then be entered into the office system for document storage and retrieval, as well as for subsequent output and distribution.

Voice Systems

Voice systems traditionally involve two steps: dictation and transcription. Dictation systems vary widely but normally involve speaking into a portable or stationary recording device whereby the system information is recorded magnetically on cassette, disk, or a similar medium. With telephone dictation systems, a telephone number is dialed (from anywhere in the world), an access code is given, and the user begins to speak. Most dictation systems permit the speaker to review and revise the input and include instructions for the person who will play the message back.

Transcription involves playing back the recorded message for typing or for entry into a word processing system.

In a *voice distribution system,* voice messages are stored in a digital form in a computer located centrally. Thus a message to be distributed is entered and a receiving party is identified. The voice distribution system, consisting of special hardware and software, takes care of the distribution process whereby a recipient of a message can receive a playback of the messages stored under his or her user identification code. From there, a recipient can respond to the message by supplementing the existing voice recording and further distribute the updated message or reply to the sender.

Facilities also currently exist for combining textual and voice information. The "written" word in this case is annotated by recording signals, usually represented by dots on the screen corresponding to a fixed duration of speech, such as a second. When the screen's cursor is moved past a recording signal, an audio unit is activated to play back the corresponding voice. Appropriate entry, storage, and distribution facilities are provided through special hardware and software.

Reprographics

Reprographics is a general term referring to the whole class of output functions mentioned previously. The usual operations are:

Printing.
Copying and duplicating.
Collating, folding, and binding.
Addressing and packing.
Platemaking and printing.

The various functions are represented collectively because of the advent of the *smart copier,* which can combine printing, copying, duplicating, and collating into the working of a single machine by adding hardware and software facilities for processing, storage, and telecommunications. Machines in this class provide a wide variety of typographic options and work in conjunction with other office automation systems, such as word processing and automatic mailing equipment.

Electronic Mail Systems

Two aspects of communication in an office environment that are age-old problems are transport delay and availability. *Transport delay* simply refers to the time necessary to transmit information from one location to another, regardless of the medium. The postal system, in-house mail, and specialized carriers all require time to move information, even under the best conditions. *Availability* refers to the obvious situation that when two or more people need to communicate under ordinary circumstances, they must be available to each other. Clearly, this fact applies to what is known as "synchronous" communication, such as face-to-face meetings or telephone conversations. The phenomena of telephone tag and of people working in different time zones, both mentioned earlier, are obvious cases of unavailability.

Electronic mail systems allow a direct attack on the problems of transport delay and availability. However, not all achievements in this important area are independent of other office automation functions, and others represent relatively mature technology.

In a modern office setting, one of the most far-reaching applications of electronic mail is represented by the following scenario, typical of everyday operations:

- Two or more workstations with word processing capability are connected by data communications facilities to a remote computer that is installed to serve the workstations in an office environment.
- A document is prepared at one workstation for review by a person at another workstation and for a signature by a person associated with a third workstation.

- The document is prepared at the originator's workstation, through retrieval and word processing facilities.
- The document is stored in the remote computer so that it is accessible by another workstation, perhaps some distance away.
- At the second workstation, the document is reviewed and possibly edited.
- At the third workstation, the modified document is printed and signed for dispatching.

This is a typical series of events that might take place in an office of a contracting or engineering firm. A model of a contract is retrieved from electronic storage by the originator and modified for the present case by inserting text and possibly numerical information. After completion, a telephone call is made to the legal department to say that the document is available for review via the computer. The document is okayed there after conclusion of a paragraph, such as a disclaimer. The originator now has a final document. The president's office is then informed that the contract is available for handling. Only then is the contract printed by the president's information administrator for signature and mailing. This application clearly reduces the time necessary for internal mail delivery and completely eliminates the need for draft copies and retyping. While the personnel involved operate at the same level of functionality, the actual activities performed exist at a higher level of sophistication in the sense that a support worker becomes more of a knowledge worker.

Another common application involves a field sales force where salespeople are provided portable computer terminals. The salesperson meets with a client to discuss particulars of a pending order. Specifications and requested delivery dates are transmitted to the home office as a message file while the customer is taken out to lunch. At the home office, an information administrator interrogates the message file and provides the requested data to the salesperson's personal file. After lunch, the salesperson checks the message file and closes the deal. Finally, a message is sent to the home office to start production, with the indication that "the paperwork is in the mail."

A third application is concerned with executive work and the availability of participants in the communication process. The handling of routine messages is commonly regarded as a time waster and in many cases is very inefficient. For example, a five-minute call may involve four minutes of small talk for one minute of business communication. In other cases, normal telephone interruptions can be disruptive of thought processes and certainly not manageable on a priority basis. In still other cases, some people are simply too difficult to deal with and it is easier and more convenient to send them a message for an expected brief reply. This application is called an *electronic message system,* and it is what most people regard as electronic mail. Each participant has an electronic mailbox in a central or remote computer that serves as a repository for messages. In most systems, the electronic mailbox has incoming, outgoing, and active components and provides a facility for sending to a distribution list and for conveniently appending a simple reply to a message and returning it. In many cases, a portable terminal is available for a manager to take home, reducing the need for late-night and weekend work. Electronic message systems are also useful for people who travel. Electronic messages are *asynchronous communications,* since the parties in the process are not active at the same time.

The second and third applications covered are properly classed as *computer-based message systems.* The first application is frequently called *communicating word processors.* Other types of electronic mail in common use are facsimile and Telex (or TWX) message switching.

Conferencing

Recently, a chain of hotels in the United States announced a videoconferencing facility throughout its hotel system. This advanced technology service will link two or more groups of people through closed-circuit live television transmitted by satellite. Clearly, the service is intended for participants who cannot be in the same city at the same time. Videoconferencing is representative of options that are available for conferencing in the modern office.

Most advances in conferencing are an outgrowth of *audiocon-ferencing* or, as is commonly known, the "conference call." Audioconferencing consists of three or more people in two or more locations involved in a process of simultaneous communication through the use of microphones and speakers connected through a telephone network. There is considerable saving in travel and personnel productivity inherent in audioconferencing; however, the lack of face-to-face interaction and informational commonality are distinct disadvantages of this level of technology.

One level up from audioconferencing in a modern office environment with communicating word processors is suggested by the following dialog:

Knowledge Worker 1: "Put the Ace report on the screen and let's discuss it."
Knowledge Worker 2: "O.K. What's it stored under?"
Knowledge Worker 1: "ACE32W."
Knowledge Worker 2: "I've got it. Now, what is it that you want to talk about?"

The situation represents a case where a document is stored in a computer and is accessible to several workstations. The advantage of this approach is that a normal telephone conversation is supplemented by a centralized information resource. This application is sometimes called "teleconferencing."

Videoconferencing combines TV facilities through satellite, cable, or leased telephone lines with the benefits of audioconferencing. Thus nonverbal communication is possible, making this method applicable to marketing presentations, training sessions, fund raising, seminars, shareholders' meetings, and so forth. Videoconferencing can be augmented with a technique known as *compression*, in which the required transmission bandwidth is reduced by not retransmitting portions of an image that have not changed since the previous transmission. Compression is commonly referred to as the "freeze frame" mode, for obvious reasons.

The term *computer conferencing* refers to the use of computers and data communications equipment for conducting an on-line conference among parties in remote locations without requiring face-to-face and synchronous activity. In most cases, a computer conference has a chairman, an agenda, and some sort of time domain. Messages are sent between participants in an asynchronous mode; at the same time, the messages are logged under the conference title. A primary advantage of this technique is that it permits a response to be prepared carefully through research or local consultations before it is entered into the system.

Thus far, computer conferencing is expected to be particularly useful in two instances: working at home and technological forecasting. Working at home through the use of office automation facilities is characteristically called *telecommuting*, because, for many types of activities, it eliminates the need for coming into the office. Although telecommuting is closely akin to the use of electronic message systems and may in fact use the same computer and communications facilities, it differs in that it is production-oriented rather than communication-oriented.

Computer conferencing is also used for *technological forecasting* to obtain a consensus on a topic among a panel of experts. Through computer conferencing facilities, a judgment is requested from experts together with supporting evidence. Results are tabulated, and a summary and profile of responses is returned to the participants. The experts are requested to reevaluate their judgment on the basis of accumulated evidence. The process is repeated until a form of consensus is achieved.

In all the cases cited, the use of automated conferencing facilities reduces the pressures on individuals to come up with a quick response and permits any given person to participate in several task forces at the same time.

Summary

Office automation can be viewed as the process of using people, procedures, and equipment within an office environment. Because office automation is a set of procedures together with appropriate equipment, it is useful to categorize the functions that are performed:

Entry and generation.
Transport.
Storage.
Retrieval and query.
Output and distribution.

Each of these functions plays an important part in one or more of the traditional office processes, summarized collectively as planning, organizing, and controlling.

The major areas of office automation are word processing, optical character recognition, voice systems, reprographics, electronic mail systems, and conferencing. The term *word processing* is a general way of referring to text preparation through the use of a computer or its equivalent. *Optical character recognition* (OCR) refers to the physical process of reading documents prepared on conventional office typewriters for entry to and storage in a computer system. A *voice system* covers the traditional office functions of dictation and transcription together with modern methods of storage, retrieval, distribution, and messaging. *Reprographics* in general refers to the whole class of output functions that includes printing; copying and duplicating; collating, folding, and binding; addressing and packing; and platemaking and printing. *Electronic mail systems* avoid the age-old problems of transport delay and availability through data communications and computer facilities. This automation area also includes mature technology, such as facsimile and Telex (or TWX). *Conferencing* includes audio conferencing, video conferencing, teleconferencing, and computer conferencing and is based on computer and communications technology. It attempts to solve the problem of travel and scheduling in a modern office environment.

There are other aspects to the modern office, but the topics given here cover the major areas. Other dimensions of office automation are covered in later chapters.

SUGGESTED READING

Saffady, W., *The Automated Office: An Introduction to the Technology*, Silver Spring, MD: National Micrographics Association, 1981.

Thurber, K. J., *Tutorial: Office Automation Systems*, Los Alamitos, CA: IEEE Computer Society Press, 1980.

Uhlig, R. P., D. J. Farber, and J. H. Bair, *The Office of the Future: Communication and Computers*, New York: North-Holland, 1979.

Zarrella, J., *Word Processing and Text Editing*, Suisun City, CA: Microcomputer Applications, 1980.

PART II

Basic Concepts

CHAPTER 3

Computers

The traditional view of a computer is that it is a numerical transformation machine capable of doing calculations at high speeds; in fact, a computer is commonly regarded as a sort of super calculator. In that role, three activities are of concern: input, processing, and output. In the realm of office automation, it is already recognizable that the architecture of the office computer takes the same form as it does with conventional data processing. Actually, the same kind of computer can be used for both, because the computer can handle nonnumeric data, such as symbols and words, in addition to numerical information.

Basic Computer Concepts

Even though a computer is commonly regarded as a black box by people who must use one, it is necessary to look below the cover to analyze its basic functions. The three basic activities are delineated as:

• *Input*—the activities whereby data on which the computer is to operate are entered into the machine. The input can originate, as far as the computer is concerned, from a keyboard device, a storage medium such as disk or tape, or any other special input mechanism.

• *Processing*—the activities whereby the data are manipulated to meet the needs of a given application. Processing may be simple, as in the case of retrieving information for the user, or more complicated, as in the case of analyzing and comparing various categories of information.

• *Output*—the activities whereby the results are made available for further use. Output may involve display information on a screen, generating a report, storing information on mass storage devices, or transmitting information to another location.

In order to support a set of computer applications it is necessary to have hardware units that support these three activities and at the same time permit the computer to operate automatically. Here, "automatically" means that the computer must be able to go from one instruction to the next and from one activity to the next without operator intervention.

In this day and age, it is no revelation to say that a computer operates under control of a series of "computer instructions" called a computer program. The manner in which a computer handles instruction processing is similar to the way a person would solve a problem using pencil, paper, calculator, and a list of instructions. The instructions and data reside on the paper, and the calculations are performed through the use of the calculator under control of the person. Reading and writing correspond to input and output, respectively.

An automatic computer has the following components:

• A *memory unit* for holding programs and data while the computer is operating.

• A *processing unit* for manipulating data and controlling the manipulation process.

• A *micro storage unit* for holding programs and data for long periods of time, but providing fast access to information when necessary.

• A *keyboard/display unit* for permitting an operator to communicate directly with the computer and control its operation.

• *Input and output units* concerned with reprographics, optical character recognition, data communications, and so forth.

Collectively, these components constitute a system like the one shown in Figure 3-1. During the execution of a program, the various components operate together by exchanging control signals and data.

Each class of component is covered in succeeding sections. Keyboard/display units are grouped under the category of input and output.

Memory Unit

The memory unit of the computer serves as an adjunct to the processing unit and is used to hold a program while it is being executed, together with associated data. These days, memory is synthesized from electronic components and imbedded on a chip. Two other commonly heard names for memory are "main storage" and "main memory." Care should be taken not to confuse memory with a mass storage device, such as magnetic disk. The two are used for different purposes.

Without going into detail about how memory is used and how it operates—which is not specifically necessary for understanding office automation—there are three important aspects of memory that must be discussed: (1) words versus bytes, (2) ROM and RAM, and (3) memory size. These factors effectively determine how satisfactory a given memory is for a specific office automation application. "Words versus bytes" refers to how the computer's memory is organized. The concepts of read-only memory (ROM) and random-access memory (RAM) determine how memory is used and, in effect, how the computer operates. Memory size is concerned with how much memory is needed to sustain an office automation function.

Bytes and Words

Memory is composed of locations that are assigned numerical addresses through which the contents of the locations can be referenced. What is of concern here is the amount of data that a location can hold. Two general concepts for organizing memory are in widespread use: bytes and words. A *byte* is a unit of storage that can hold eight bits of information. In these eight bits, a single charac-

Figure 3-1. Schematic diagram of a typical computer system.

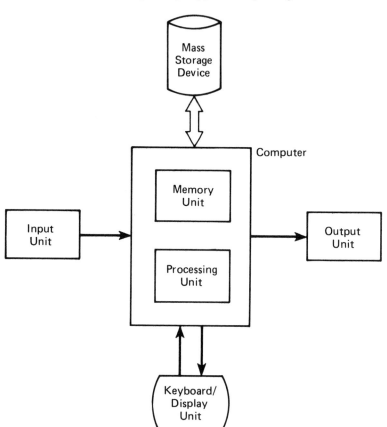

ter's worth of information can be represented or stored, as the case may be, or the eight bits can be part of a longer string of bits needed to represent a number. In a *byte-oriented computer,* each location can hold a byte. A *word* is a larger unit of storage, usually 16, 24, 32, 36, or even 60 bits. Word sizes of 16 and 24 bits are common. In a word, several characters' worth of information or a numerical value can be represented or stored. In a *word-oriented computer,* each location can hold a word.

It is common in office automation circles to discuss how much memory is needed to support a given application, and it is well to note that there is a very significant difference, for example, between 128,000 bytes and 128,000 words.

Read-Only Memory

Computer memory is divided into two classes: ROM and RAM. Read-only memory (ROM) is used to hold programs and small amounts of data that do not change during the course of computer operation. The information that is held in ROM is "burned in" when the computer is assembled and cannot be modified by the user. As an example of how ROM is used, consider a microcomputer that responds immediately when the power is turned on by "READY" on the screen. This happens because the computer is under control of instructions held in ROM. Any data held in ROM deal with the parameters of the computer system.

As far as the use of ROM in office automation is concerned, several possibilities exist:

1. No ROM is used. When the computer is turned on, it is "dead" and all instructions must be "bootstrapped" from disk or tape. (Bootstrapping refers to a process whereby a program can load itself into memory.)

2. ROM is used for error checking of the computer circuits. When the computer is turned on, it runs through a self-check sequence of the instructions held in ROM. Then computer instructions must be loaded in before any productive work can be done.

3. Programs in ROM take care of error checking and screen handling. In this case, the error checking is taken care of as described above, but additionally, ROM contains programs for doing screen editing, scrolling, and data transfer between computers. As above, applications programs, if they exist at all, must be loaded into memory for execution.

4. ROM contains more extensive programs for managing the operation of the computer and providing facilities for creating computer programs. In this case, the user's programs can be conveniently written, stored, loaded, and executed through the computer instructions held in ROM without additional programs provided for the computer having to be loaded in.

The use of read-only memory is important in office automation, and most existing computer hardware and workstations reflect one of these cases.

When the power to the computer is turned off, the contents of
ROM remain intact. Thus, when the computer is turned on, it can
respond without delay.

Random-Access Memory

Random-access memory (RAM) is also used for holding programs and data, but it can be read and written under program
control. Information placed in RAM is held there until the corresponding location is rewritten or until the power to the computer is
turned off, at which time all information in RAM is lost. When a
program is loaded, for example, it is read into the computer and
placed in RAM.

The term "random access" gives some indication of how this
type of memory operates. With all random-access memory—and
this includes both RAM and ROM—the speed with which a location can be accessed is independent of its position, making it an
efficient mechanism for storing information within the computer.

Without going into too much detail, it is useful to consider a
simple case of the kind of information stored in RAM. Assume a
computer has a screen of 24 lines, each containing 80 characters. A
screen buffer of 1,920 characters (that is, 24 lines times 80 characters) is needed in RAM to hold the contents of the screen. With
byte storage, the buffer would require 1,920 bytes. With 32-bit
word storage at four bytes per word, the buffer would require 480
words.

Memory Size

The amount of memory needed in a computer system is dependent on the application domain. If a single office automation
function is performed in a computer, then the amount of memory
required is usually dependent on the options that are available. If
several office automation functions are performed in the same
computer, a careful analysis is needed to configure a viable system.

Word processing serves as a realistic example. A standard word processing system that permits creating, storing, retrieving, formatting, and printing textual information may require 128,000 bytes of memory on a typical computer. A semiautomated text processing system with the capability of merging portions of several documents, such as would be done in synthesizing a contractual agreement, could require 196,000 bytes of memory. Finally, a fully automated system allowing form letters to be completed with keyboard or file input, special headers and footings for pagination, global replace/delete operations, and automatic spelling and hyphenation may require as many as 256,000 bytes of memory. This memory capacity is used in two ways: (1) to hold the computer programs providing the given functions and (2) to hold more text so that fewer operations require that information be read and written to disk storage. Thus bigger memory improves efficiency up to a point because it decreases the number of input and output operations, which take time and can be frustrating to a support worker.

Normally, the vendor specifies the memory requirements for a given application—say, word processing package I, x units of memory, word processing package II, y units of memory; adding automatic spelling and hyphenation to either of them requires an additional m units.

Processing Unit

The processing unit is the key element in the computer system, because it controls the operation of other units and executes the programs. The processing unit operates by fetching instructions from either ROM or RAM, interpreting them, and then executing the operations specified in the statements. The processing unit is commonly referred to as the central processing unit (CPU) or the microprocessor. As an example of how and when the processing unit shows its presence, consider the case of a rather long document held in RAM and the execution of a word processing function, such as a global replacement, that is nontrivial in the sense that several seconds of processing time are ordinarily required for it. If the processing unit is fast, the function will take only a few seconds—in fact, the operator may not even be able to gauge the time. On the other hand, if the processing unit is relatively slow, even the operator may be able to detect the slowness. The time may be compounded by the size of RAM. If RAM is so small that the document has to be read into it in pieces for processing, the time necesary to do the function will necessarily be longer than if the whole document can fit into RAM at one time.

Organization of the Processing Unit

The processing unit comprises two components: a *control unit* and an *arithmetic logic unit.* The control unit provides the capability for automatic execution by going from one instruction to the next without operator intervention. The control unit operates by fetching an instruction from either ROM or RAM and interpreting what computer function it specifies. Then the control unit sends signals to an arithmetic logic unit to actually do that computer operation.

Almost without exception, a processing unit can perform text manipulation, arithmetic, and a variety of communications functions. Precisely how a processing unit is used is up to the designer of an office automation system.

Speed of the Processing Unit

Most processing units are fast enough to do office automation work. In fact, there is rarely an occasion to even be concerned with overall speed, except for a few cases introduced below. Even most microcomputers available at the local computer store have more than enough processing power to do office automation for a single user. Some microcomputers incorporate the same microprocessors as those used by office automation equipment available from large computer vendors.

Speed of the processing unit is critical in *shared-logic systems.* A shared-logic system is one in which the computer is sharing its processing power among several office automation users or a combination of office automation, data processing, or data communications requirements. The problem is obvious. The computer becomes so loaded doing everybody's work that no one really gets good service. This topic will be covered in more detail later; however, the solution is as obvious as the problem. If a system does become saturated, the bottleneck can be relieved by buying a faster processing unit or reducing the number of concurrent users.

Intelligence

There are several places in an office automation system where a processing unit can be found. In a stand-alone word processing unit, the equipment must contain memory and a processing unit, plus other devices. This is obvious because the processing unit is needed to do the work.

In systems where two or more workstations are connected to a computer, the computer for sure will contain memory and a processing unit, plus other devices as in the stand-alone unit. Clearly, a more powerful system is needed to service several users than to service a single user, but the overall concept is the same. A workstation itself may or may not contain a processing unit. If the design of a workstation incorporates a processing unit, it is said to have intelligence—hence the name *intelligent terminal* or *intelligent workstation*. If the design of a workstation does not incorporate a processing unit, it is said to be dumb—hence the name *dumb terminal*.

An office automation system with intelligent workstations is not necessarily better than a system with dumb workstations. "The system is the solution," so whether a system is appropriate for a given office depends on overall functionality rather than the characteristics of a particular box.

Mass Storage Units

The effectiveness of a computer system is directly related to its capacity for storing information—that is, programs, textual information, numerical data, images, voice data, and so forth—over long periods of time. Two types of mass storage devices are customarily used: magnetic tape and magnetic disk. Other technologies exist, but tape and disk are in common use.

Tape

Magnetic tape is a *serial medium,* which means that the computer must pass over all preceding elements before a given unit of information is reached. Cassette tape and conventional ½″ computer tape are the most common; cassette tape is commonly used with microcomputers, and magnetic tape is used with medium- to large-scale computers. Magnetic tape is infrequently used with office automation.

Disk

Magnetic disk is a *rotating medium* that permits direct access to a unit of information. A magnetic disk is a disk coated with magnetic recording material; the concept is similar to a phonographic record because data are recorded on tracks and are read or written as the disk rotates. There are some notable differences. A phonograph record has grooves and a magnetic disk does not; also, the tracks on a disk are concentric, whereas the grooves on a phonograph record are spiral.

Disk storage is available in two forms: hard disk and soft disk. The term *hard disk* denotes a set of metal disks, coated with magnetic recording material and mounted on a spindle. Each disk is approximately 14″ in diameter and is usually removable. Tracks are accessible via read-write heads attached to a control arm that moves in and out. A disk access is performed by moving the read-write heads to a specific track address—called direct access—and by performing the requested read or write operation. If the read-write heads, control arms, and disk assembly are constructed in a sealed assembly, the removable unit is called a *Winchester disk.* Some disk drives use fixed units that cannot be removed, and others use removable disk packs where the read-write heads are permanently attached to the disk housing. Figure 3-2 gives a schematic of a hard-disk unit.

Figure 3-2. Schematic diagram of a hard-disk drive.

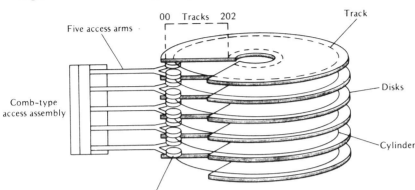

From Harry Katzan, Jr., Introduction to Computers and Data Processing (New York: D. Van Nostrand Company, 1979), p. 135. © 1979 by Litton Educational Publishing, Inc. Reprinted by permission of Kent Publishing Company, a division of Wadsworth, Inc., 20 Park Plaza, Boston, MA 02116.*

The term *soft disk* refers to a flexible disk storage medium, normally called a *diskette* or *floppy disk.* "Floppy" suggests the fact that the diskette wobbles as it rotates because of its flexible construction. A diskette is either an 8″ or a 5″ circular piece of flat Mylar® polyester sheathed in a polyvinyl chloride protective jacket, as shown in Figure 3-3. When a read or write operation is performed on a diskette, the Mylar® disk rotates in its protective jacket.

Eight-inch diskettes are commonly used with small business computers and professional word processing and other office automation systems. Five-inch diskettes are most frequently used with personal computers.

Diskettes are particularly appropriate to small-scale operations because of the relative ease with which they can be handled and stored. For large-scale operations, hard disks are preferable because of faster access and transfer rates and large storage capacity.

Figure 3-3. Schematic diagram of a diskette. (*Courtesy Petrocelli Books, Inc.*)

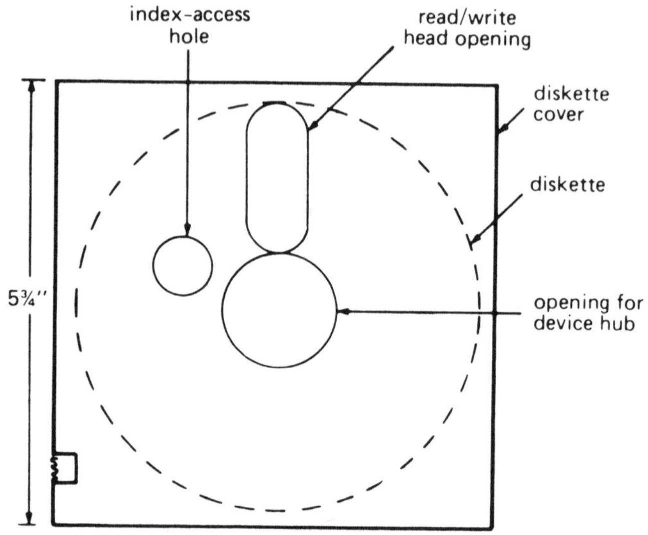

Input and Output Units

Input and output units are used to transfer information between a computer and the outside world. Devices in this category
range from card readers to optical character readers and from
printers to audio output units. Three classes of input and output
devices are particularly important to office automation: the keyboard, the display, and the printer. At least one of these units is
found in almost all office automation systems, regardless of the automated functions and the applications domain. Figure 3-4 gives a
picture of a typical automated office workstation. (The unit shown
is in fact a stand-alone word processing system, with the processing
unit located in the box below the screen.)

Keyboard

The keyboard is an integral part of most office automation systems and resembles an ordinary office typewriter in function and

Figure 3-4. Typical automated office workstation with printer. (*Courtesy Datapro Research Corporation.*)

in appearance. It is the user's primary means of entering information into the system. Figure 3-5 gives a schematic of a typical keyboard, characterized as follows:

- A standard QWERTY structure similar to a conventional typewriter.
- Special-function keys located to the right and to the left of the main key group.

The special-function keys permit operations such as cursor movement, scrolling, text manipulation, and workstation control.

Display

The keyboard and the display unit operate in tandem in office automation workstations. Characters entered through the keyboard are shown on the display, such as the unit depicted in Figure 3-6. The design of most modern display units has gone through an evolution of human-factors research in order to present a product that is responsive to the needs of office workers. Some of the recent enhancements in screen design are:

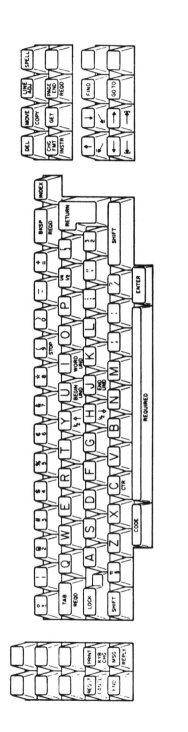

Figure 3-5. Schematic diagram of a typical keyboard. (*Courtesy International Business Machines Corporation.*)

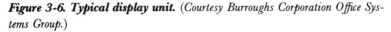

Figure 3-6. Typical display unit. (*Courtesy Burroughs Corporation Office Systems Group.*)

- Green phosphor screens to reduce eye fatigue.
- Antiglare etchings and/or specially coated screen surfaces for better operator performance.
- Tilted screens.
- Optimum character height, width, and spacings.

Research continues in this important area, and the effective office manager can improve productivity by staying abreast of current developments.

The format of the screen is of great significance for word processing, since document design is still a human art. A screen should be wide enough and deep enough so that "what you see is what you get." The information assistant should be able to visualize a page on the screen as it will look when it is printed. Figure 3-7 gives a typical display format for word processing composed of 24 lines, each of 80 characters. It is possible to get by with fewer lines per

Figure 3-7. Typical display format. (*Courtesy International Business Machines Corporation.*)

```
 1 Context Field          IDocument Name                IAudit Window    I
 2 L DiskIR DiskICommunication StatusIPg xxxx.y.zILn zxxxyIKyb xxx-yIPitch xx
 3 ....:....1....:....2....:....3....:....4....:....5....:....6....:....7....8
 4
 5
 6
 7
 8
 9
10
11
12
13
14
15
16
17
18
19
20
21
23
24 prompt
25 message
```

screen and also fewer characters per line. However, this decision must be made on a case-by-case basis.

Printer

A printer in office automation terminology is a device for generating hard copy in the form of letters, reports, documents, and draft copies. Figure 3-8 shows a typical printer with an automatic paper feeder. Printers are, in general, classified by three criteria: paper, printer mode, and printing mechanism.

Paper can be fan-fold, roll, or cut sheets. Fan-fold paper is the conventional continuous computer paper. Roll paper is also continuous but without perforations for tearing. Cut sheets are letter-quality stationery. With cut sheets, automatic paper feeders are available for some printers, eliminating the need to insert individual sheets.

Printer mode refers to whether the printer prints a character at a time or a line at a time. Most office systems use character printers with speeds that range from roughly 40 to 200 characters per second (cps).

Figure 3-8. Printer with paper feeder. (*Courtesy Datapro Research Corporation.*)

Printing mechanism refers to whether the printer uses an impact or nonimpact mechanism. Nonimpact methods use laser beams, ink jets, xerography, or electrostatic means. Impact character printers use a dot matrix, a daisy wheel, or a rotating cylinder or sphere. Characters are formed with a *dot-matrix printer* as a rectangular arrangement of dots. Typical dot matrixes are 7 × 9 and 5 × 8. With a *daisywheel printer,* characters are embossed on a wheel with rays emanating from the center. At the end of each ray is a chacter embossed on a slug—hence the name "daisywheel" or "printwheel." The daisy wheel rotates, and a hammer strikes the

Figure 3-9. Typical universal terminal. (Courtesy Wang Laboratories, Inc.)

daisywheel when the slug is in the correct position. The rotating-cylinder or -sphere concept uses characters embossed on the print element, which is rotated electronically prior to impact. All character impact printers use a print head that moves across the page and a ribbon to provide the hard-copy image.

Daisywheel (or printwheel) printers are generally used for better-quality work, and matrix printers are commonly used for draft copies. However, many matrix printers yield high-quality output, and customers frequently regard them as being of "letter quality."

The Universal Workstation

Many offices already contain terminals for specific applications such as data entry, banking, word processing, and so forth. From an organizational point of view, it is desirable to use terminals or workstations for multiple purposes, since an operator is thereby provided with multiple resources in the same work domain.

The "pie in the sky" of office automation is a *universal workstation* capable of serving several needs of the organization. Some of the characteristics of a universal terminal, such as the one depicted in Figure 3-9, are that it is microprocessor-based and has an ergonomically designed keyboard and display screen. Because of the microprocessor basis, the unit can serve as a terminal to a remote computer, a stand-alone word processor, or a small business computer.

Summary

The modern view of a computer is that it is a machine that can handle numeric and nonnumeric information through the execution of three basic activities: input, processing, and output. Hardware units support each of these activities and provide for automatic execution of computer programs.

Automatic computers have the following components: memory units, processing units, mass storage units, keyboard/display units, and input and output units. The various components work together by exchanging control signals and data.

The memory of a computer is used to hold programs and data during execution and is characterized by memory unit (words or bytes), memory type (read-only memory or random-access memory), and memory size.

The processing unit controls the organization of a computer system by fetching instructions from the memory unit and executing them. The processing unit is composed of two major elements: the control unit and the arithmetic/logic unit. The speed of the processing unit together with the size of memory determines the effectiveness of a computer system for office automation applications. Workstations that contain processing units are called "intelligent" terminals or workstations.

The major types of mass storage devices are magnetic tape and magnetic disk. Magnetic tape is a serial device and is infrequently used in office automation. Magnetic disk is used extensively for storing programs and data, including textual, image, and other forms of information. Magnetic disk comes in two varieties: hard disk and soft disk. Hard disk refers to a set of metal disks coated with magnetic recording material and mounted on a spindle. Hard disks are commonly used with large-scale systems. Soft disks, known as diskettes or floppy disks, are flexible and used for smaller-scale operations.

Although input and output devices of every imaginable type and variety are used with office automation systems, the three most important classes of input and output devices are the keyboard, the display, and the printer. Collectively, a set of these devices is termed a workstation. A universal workstation is one designed for multiple functions in an office environment.

SUGGESTED READING

Chorafas, D. N., *Office Automation: The Productivity Challenge*, Englewood Cliffs, NJ: Prentice-Hall, 1982.

Fohl, M. E., *A Microprocessor Course*, Princeton, NJ: Petrocelli Books, 1979.

Katzan, H., *Invitation to FORTH*, Princeton, NJ: Petrocelli Books, 1981 (Chapter 1: "Computer Fundamentals," pp. 13–34).

CHAPTER 4

Software

The term *software* is used in many industries to denote "that which is not hard, like electronic equipment or other physical machines." Some people say you can paint hardware red, but not software. In some businesses, rules and procedures—sometimes called programming—are known as software, and so it is in the fields of computers and office automation.

The objective of this chapter is to present an idea of what software is without going into needless detail. Here is a basic example to set the stage. When you buy a stand-alone word processing system, you receive several pieces of equipment: keyboard, display, control unit (that is, the box under the display), disk unit, and a printer. The control part contains the computer. The units may be assembled together in some fashion. For example, the control unit may be located with the keyboard or the display, or the control, keyboard, and display may be one unit. The disk unit is normally self-contained, but it may also be packaged with some of the other units. The printer is normally a self-contained device. Not to belabor the point, this is the hardware. The programs that drive the word processing operation are held in ROM or reside on diskette or hard disk. These programs are the software.

A key advantage to using software, as compared to hardwired electrical components, is that software is more tractable. It can easily be changed. A user can upgrade from word processing package I to word processing package II simply by using another diskette. The vendor can easily distribute software in this form.

Software that is specially oriented to the particular computer can be held in ROM or on disk, or a combination of the two. On the other hand, software that has an application orientation is best stored on disk, so that it can easily be changed and new applications can conveniently be added.

Traditional Computer Applications

Computer applications are usually grouped into some well-defined classes that can be given a unique name and succinctly described. In reality, most applications span one or more of the classes, since many enterprises will combine office automation with existing computer applications. Overall, however, the classes give a pretty good idea of what is being done with computers.

Data Processing

Data processing is the storing, processing, and reporting of information for an enterprise. Data processing is uniquely concerned with record-keeping functions, such as those involving accounting operations. These functions have the following characteristics:

High volume of input and output.
Relatively little processing per record.
Fixed record and filing structures.

Most computer applications that have these characteristics, such as billing, transaction processing, payroll, check processing, and inventory control, are classed as data processing.

A data processing workload is characterized by periodic computer runs.

Problem Solving

Problem solving is traditionally associated with scientific and engineering work in which calculations are performed in support of design and analysis activities. A typical example would be the calculation of design parameters for a structure, such as a building, road, or bridge. Calculations for business analysis also fall into this broad classification. Problem solving is characterized by a small amount of input, a relatively large number of calculations during processing, and a small amount of output representing the results of the calculations.

Problems in this category vary in scope and magnitude, and range from programs that compute the trajectory of a space vehicle or trace the path of a nuclear particle to straightforward data analysis and business forecasting.

A problem-solving workload is characterized by periodic runs and a high amount of machine computation.

Computer Models and Simulation

A computer model is an abstraction of a real system, where the important variables are simulated to allow conclusions to be drawn about the real system or predictions to be made about the future. Computer models normally exist for a nation's economy, a space voyage, the toll booths for a highway, waiting lines in a bank, and the flow of data in an office automation system, to name only a few examples.

A computer model is a program that runs on a computer and is similar in concept to problem solving, except that it is more extensive and frequently uses software developed specifically for modeling.

Feedback and Control Systems

A feedback and control system enables a computer connected to a physical process to control the process in a manner and degree that is usually not possible with manual methods. This class of computer application is commonplace in laboratories for data acquisition and in many areas of the chemical-processing industry. Computer control systems in automobiles are an everyday example of a feedback and control system.

On-Line and Real-Time Systems

An on-line system is one in which a device external to the computer—such as a terminal or workstation—can communicate with a central computer and receive a response in real time. Airline reservation and banking systems ordinarily fall into this category. Normally, telecommunications facilities are used with on-line systems, and the overall system is designed to provide a prespecified level of response time.

On-line and real-time systems are frequently combined with data processing or information systems within the domain of a central computer complex.

Information Systems

An information system is a central repository of information, together with computer facilities that permit the information to be accessed. Most information systems are implemented together with on-line facilities, but this need not be the case in all systems. The objective of an information system is to provide data for data analysis and reports, for information retrieval, and for inclusion in other documents.

Frequently, special computer languages are used to reference an information system so that special computer training, other than knowledge of the specific language itself, is not required of the people involved. Languages of this type are particularly appropriate for the office environment, since they generally provide access to the same repository of information as office automation facilities that give the capability for the automatic integration of textual information and data.

Artificial Intelligence

Artificial intelligence refers to the class of computer applications that perform tasks that are regarded as intelligent when done by humans. While this area is historically associated with game playing, theorem proving, and robots, a modern view of the subject matter would necessarily include voice recognition and linguistic analysis. In the latter category, some information systems provide natural-language inquiry facilities that clearly place them in the office domain.

One way of looking at artificial intelligence is that it provides a means of doing not only esoteric machine processing but also jobs that are simply too dull for human beings.

Office Automation

It is not surprising to see office automation listed as a major computer application. However, it is generally felt that office automation, as introduced earlier, can affect the total enterprise and spans most of the other classes of applications. Because office automation is a set of procedures together with appropriate equipment, it is as much a concept as it is a computer application. The notion of integrating data processing, data communications, and office automation into a total system effectively expands the traditional view of data processing into an "information business" totally contained within the organization.

Distributed Information Systems

In a distributed information system, the information resources of an enterprise are distributed across several organizational units. Figure 4-1 gives an overview of the layers of components in a distributed information system. In most systems of this sort, the various centers have an analogous structure without necessarily being identical.

At the center of the diagram is the computer system, which provides the "power" of the distributed information system. Proceeding in a layered fashion, the software of the system consists of the systems software, applications subsystems, and the applications programs. The center of the diagram corresponds to the most critical component in the diagram; the technology is very complex and few people are involved. As one moves outward from the center, the technology becomes less complex but more people are personally affected by the system. At the outermost layer, people using the computer need not be computer specialists.

The computer system components and the application programs were covered previously. The rest of this section and other sections cover the systems software and the applications subsystems.

The *systems software* is the set of computer programs that manage the hardware units and provide the link to the other software components. The *applications subsystems* are the interface between

Figure 4-1. Layers of components in a distributed information system.

Application Programs
Data Processing
Information Systems
Office Automation, etc.

Applications Subsystems
Data Base Systems
Transaction Processing Systems
Inquiry/Response Systems

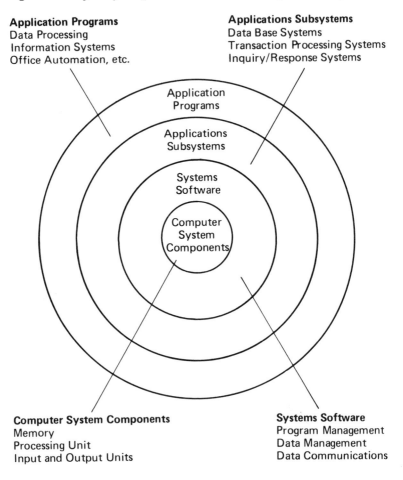

Computer System Components
Memory
Processing Unit
Input and Output Units

Systems Software
Program Management
Data Management
Data Communications

the resources of the system and the user, in the case of on-line systems, or other applications programs. The resources of the system are the hardware units, the facilities of the systems software, and the information stored in the system. The user is the operator at a workstation or terminal, as in the case of an airline reservation system.

Systems Software

Systems software, as introduced above, provides four major functions in the total computer environment:

1. An interface between the hardware and the other software in the system.
2. An interface between a user at a terminal or workstation, or even an application program, and the software.
3. An interface between a user and the information resources of the system.
4. A means of communicating with remote stations.

Not all instances of systems software contain facilities for each of these functions. However, even the most elementary system contains the first and second items.

Monitors

If the systems software is contained only in ROM, it is called a *monitor,* which provides user facilities in a relatively small-scale environment. In effect, a monitor controls the execution of programs by supplying the following functional capabilities:

- Automatic startup of the computer from ROM when the power is turned on.
- Handling user input from the keyboard and generating the output of information on the video display.
- Loading and saving application programs from mass storage devices.
- Running and listing application programs.
- Providing the printer or mass storage devices with input and output facilities for application programs.

Monitors stored in ROM are limited by the size of ROM, so that operating enhancements to existing systems are not commonplace. Monitors are most frequently used with microcomputer-based systems.

Operating Systems

If the system software is stored on disk, it is commonly called an *operating system*. The concept of an operating system has evolved from earlier software concerned primarily with the otherwise cumbersome procedures for operating a computer.

An operating system controls the execution of application programs in its domain through three major facilities: program management, data management, and communications management.

Program management routines perform the following operations for a user of the system or for an executing program:

Loading programs from disk into RAM.
Running programs currently resident in RAM.
Saving programs from RAM onto disk.

If a program is too big to fit into RAM, it is common practice for the program to execute until it gets to a good stopping place or until a special routine is needed from disk; then it dynamically loads the needed software from disk to RAM and continues execution. This philosophy applies to both systems software and other software.

Operating system routines that are resident in RAM for the total execution time are called *resident routines*. Routines that are "loaded" only when needed are called *transient routines*. In general, there is no significant name that applies to application programs that are resident or nonresident, but the same operational philosophy applies. Thus a large word processing package, for example, may consist of a frequently used part that is in RAM for the complete execution time, such as the routines that accept textual input and do various kinds of line and screen editing. Special facilities, such as a routine to check spelling, are then loaded dynamically when the operator requests that function.

Data management routines do for data what program management routines do for programs. In many cases, program management routines actually use their data management counterpart for various input and output operations. Data management routines provide the following operational capabilities to a user of the system or to an executing program:

- Storing data files and programs on disk so they can be retrieved by name.
- Copying files from one disk to another.
- Erasing and renaming files.
- Performing disk input and output operations, thereby subordinating to systems software much of the detail normally associated with programming.

Data management facilities are also used by applications subsystems. Practically all disk operating systems incorporate data management routines. The level of complexity of an operating system is normally reflected in the sophistication of data management capability.

Data communications routines permit telecommunications to be maintained with a remote station—that is, a terminal, a workstation, or a computer—through a system of software components that perform tasks such as line control, error handling, code conversion, addressing, polling, and message construction. In a communications environment, ordinary telephone facilities are used to transmit data between remote locations, and cooperation between both hardware and software elements at each end of the data link is needed to sustain the flow. In the world of modern office automation, the use of data communications facilities is commonplace. Of course, this is also true of on-line and real-time systems and of information systems. In the area of microcomputers, communications software is now available, and in many cases, microcomputers are employed as intelligent terminals or workstations. This subject is covered further in Chapter 5.

Applications Subsystems

As an interface between the resources of the system and the user or an application program, an applications subsystem is normally designed to service several tasks on a systemwide basis. An applications subsystem is nontrivial in the sense that it would be prohibitive in cost and time for a user to develop a comparable facility individually. Some applications subsystems require special services of the systems software, while others are indistinguishable from a systems viewpoint from any other application program. Three representative classes of applications subsystems are introduced: data base systems, transaction processing systems, and inquiry/response systems.

Data Base Systems

A *data base* is a centralized collection of data organized for use by one or more computer applications or for one or more persons. In a computer facility, the computer files are integrated in the sense that redundant data are factored out. For example, assume an enterprise has both payroll and personnel files. An employee's name, address, employee identification number, and so forth are stored in both places. When a change is necessary, two files need to be updated. In a data base environment, common data are represented only once and the data unique to a given application domain—payroll or personnel in this example—are organized so that they are not in general accessible to independent applications unless permission is explicitly granted. Now when a change is required, only one update need be made.

A *data base management system* is a set of software elements that provides the capability for creating, storing, updating, modifying, accessing, copying, and otherwise managing a data base. Most data base systems use data management system software for doing mass storage operations. When an application program or another applications subsystem requires a data base operation, it links to the data base management system instead of doing its own mass storage input and output operations.

Users at a terminal or workstation interact with a data base in one of three ways: (1) through an office automation system, (2) through a specially prepared application program, or (3) through an applications subsystem for inquiry and response.

Data base facilities are particularly useful in an office environment, since automated text generation involves accessing a data base and using the retrieved information in document preparation.

Inquiry/Response Systems

An *inquiry/response system* is a set of software elements that gives a user at a terminal or workstation the capability of interacting with a data base and retrieving information in an unstructured fashion. The vast majority of software systems in this category are user-oriented, which means:

- The system is user-friendly.
- The user can interact with the system with an easy-to-learn language specially designed for query applications.
- The system can be used in a telecommunications environment.

Inquiry/response systems vary widely in scope and complexity. Two important variations are covered in the following paragraphs.

The two variations in inquiry/response systems deal with how the system is accessed and how it is referenced. The point of view is that of the user who is not a computer specialist.

As far as access is concerned, a system of this type can be menu-driven or command-driven. In a *menu-driven system,* the user is presented with a screen containing information outlining the options and also the query information that must be specified. The user simply makes selections or enters the requested information. In a *command-driven system,* the user is normally required to learn a query language designed for applications of this type. When a data base reference is desired, a simple query statement is entered. Neither of these methods is a priori better or worse than the other, but obviously, menu-driven systems are initially easier to learn and command-driven systems are inherently more powerful in the long run.

The question of menu-driven versus command-driven is evident in the design of other office automation software facilities. Some systems also provide both options to a greater or lesser degree.

Within an information store, requests can be made on the basis of indexes specially constructed for retrieval or on a contextual basis. When the *index method* is employed, each document is prefixed with a set of descriptions that can be used for the selection operation. When the *contextual method* is employed, a word or phrase is entered by the user and the inquiry/response system searches the entire document for an occurrence of the search key. The index method provides more rapid retrieval. Some inquiry/response systems provide both options.

Transaction Processing Systems

A *transaction processing system* is integrated into the systems software to some extent and permits transaction processing programs to be executed on a dynamic basis. A banking system always serves as a good example. When a bank teller enters a business transaction into a banking terminal on-line to a computerized banking system, the transaction processing software recognizes the particular business transaction and the corresponding system transactions. A typical "business" transaction might be to move money from an account for one customer to another account. Associated "system" transactions could be:

Verify identity of person making the transaction.
Check validity of request.
Verify availability of funds.
Verify existence of receiving account.
Deduct from originating account.
Make credit to receiving account.

Each independent event in the processing of a business transaction would normally be handled by a separate program. The transaction processing system receives the business transaction through the data communications system software and successively loads from disk and executes distinct programs for each of the system transactions. When the total set of operations has been completed, a response is transmitted to the banking terminal.

From the user's point of view, transaction processing is only remotely related to office automation. Transaction processing facilities may operate concurrently with office functions and data processing in a centralized computer system, and the various users may not be explicitly aware of each other's existence. In fact, all three applications environments may share a common data base. An important point, however, is that a transaction processing method may be employed as a means of implementing office automation software. Obviously, this is an overall systems design situation in which an analyst determines from a system viewpoint a viable means of implementing a given office function.

The basic idea is that each center of intelligence, such as an intelligent workstation or a distributed computer, can handle a given level of user service. Then, as service areas extend beyond the local system, a computer must pass on service requests to the next higher level, similar to the manner in which business problems are passed on to the next level of management. As with management, special transactions of this kind are handled as individual cases. Thus, a service request that extends beyond the local computer, such as an electronic message, can be sent to a regional or a central computer for forwarding. Within a regional or central computer, requests of this type can be handled, and in many systems are handled, as transactions and are processed using a transaction processing method.

Clearly, these techniques are an integral part of system design, which is covered in later chapters.

Summary

The computer programs that drive the automated office operation through the use of hardware units are collectively known as software. A key advantage to using software as compared to hard-wired components is that it is tractable and can easily be changed. Thus a user can upgrade from one level of service to another by simply changing the software.

Computer software is conveniently grouped into three classes: systems software, applications subsystems, and application programs.

Systems software is intimately involved with the operation of the computer and generally falls under monitors or operating systems. Operating systems are more powerful in scope and complexity and normally involve three major facilities: program management, data management, and communications management.

Applications subsystems provide extensive capability that is normally beyond the domain of an application program. Three categories of applications subsystems in widespread use are data base systems, inquiry/response systems, and transaction processing systems.

The class of *application programs* is not as clearly defined as the other categories of software and usually overlaps boundaries in a simple classification scheme. For purposes of illustration, traditional computer applications are data processing, problem solving, computer modeling and simulation, feedback and control systems, on-line and real-time systems, information systems, artificial intelligence, and office automation.

SUGGESTED READING

Housley, T., *Data Communications and Teleprocessing Systems*, Englewood Cliffs, NJ: Prentice-Hall, 1979.

Katzan, H., *Distributed Information Systems*, Princeton, NJ: Petrocelli Books, 1979 (Chapter 1: "Introduction and Overview," pp. 3–17).

Katzan, H., *Invitation to FORTH*, Princeton, NJ: Petrocelli Books, 1981 (Chapter 2: "Software Technology," pp. 35–56).

Katzan, H., *Operating Systems: A Pragmatic Approach*, New York: Van Nostrand Reinhold, 1973.

Sanders, D. H., *Computers in Society*, New York: McGraw-Hill, 1981.

Skees, W. D., *Computer Software for Data Communications*, Belmont, CA: Lifetime Learning Publications, 1981.

Sprowls, R. C., *Management Data Bases*, Santa Barbara, CA: Wiley/Hamilton, 1976.

CHAPTER 5

Communications

Data communications facilities provide the capability for transferring digitized data between remote locations. As introduced in previous chapters, the concept certainly is not new and exists as an integral part of most on-line, information, and office systems. What is new, however, is the degree to which data processing, communications, and office automation have been integrated in modern office systems. This chapter surveys basic communications concepts, computer networks, distributed processing, and some related office communications topics.

Overview

The age-old model of a communication system (Figure 5-1) serves as a good starting place for an overview of data communications in an office environment. In transmitting a message from one location to another, five important components are identified: a message source, an encoder, a communications channel, a decoder, and a message destination. The message source and message destination refer to the sender and receiver, respectively. The encoder puts a message in a form acceptable to the communications channel, and the decoder reverses the process. This model is so general that it applies to human communication as well as it does to computer communications.

Figure 5-1. Conceptual model of a communications system.

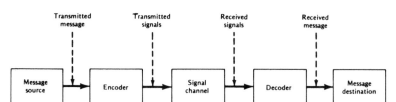

From Harry Katzan, Jr., Introduction to Computers and Data Processing *(New York: D. Van Nostrand Company, 1979), p. 145. © 1979 by Litton Educational Publishing, Inc. Reprinted by permission of Kent Publishing Company, a division of Wadsworth, Inc., 20 Park Plaza, Boston, MA 02116.*

Modulation and Demodulation

The encoding and decoding process in most forms of computer communications refers to the conversion of digital data to analog data and back again as the case may be. This is suggested in Figure 5-2. The conversion process from digital pulses to analog waves is called *modulation,* and the reverse process of converting from analog waves to digital pulses is called *demodulation.* In both cases, the conversion is performed by a hardware device called a *modem.*

Today's telephone system uses analog waves for transmission. However, this technique is not necessarily applied to all forms of computer communications.

Figure 5-2. Modulation and demodulation.

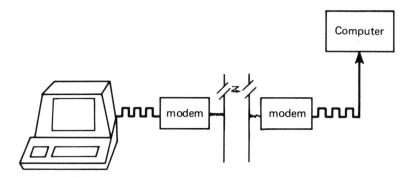

Modes of Transmission

Two modes of transmission are generally applied to data communications: asynchronous transmission and synchronous transmission. With *asynchronous transmission*, both parties need not actively participate in transmission at the same time. Electronic mail, as introduced earlier, is an asynchronous process. In data communications technology, however, the term asynchronous refers to cases where a human is in the communications loop and one character is transmitted at a time between source and destination. A character's worth of information is encapsulated in start and stop bits for calibration between transmitter and receiver. Data communication between a terminal device and a computer commonly uses asynchronous transmission because of the relative slowness of a person's response time.

When *synchronous transmission* is used in data communications, an entire block of characters is transmitted between source and destination without start and stop bits by having both parties participate in the communications process at the same time. Clearly, synchronous transmission is more efficient than its asynchronous counterpart, but it normally requires intelligence at both ends.

Office automation also has a synchronous counterpart. Videoconferencing, for example, requires the active participation of people at both ends and is regarded as a synchronous process.

Switching

A basic consideration in the design of data communications systems is the problem of connecting remote locations in an efficient manner. When direct connections are used, n locations would require $n(n-1)/2$ communication links. In the case of five centers, therefore, ten separate connections are needed, as suggested in Figure 5-3. When a switching center is used, as with the case of the telephone system, the number of connections to the switching center is equal to n, the number of locations (Figure 5-4). As depicted in Figure 5-5, a trunk line may be used for efficiency when long distances are involved.

The key point is that the notion of switching can certainly be applied to the design of office systems. Instead of having an elec-

Figure 5-3. With a direct connection, five remote locations require ten—or n(n-1)/2—communication links.

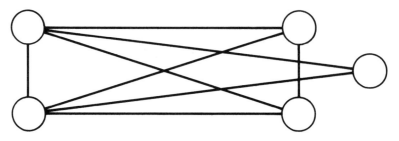

tronic mail link (for example, between every pair of workstations), it is an accepted design practice to go through a remote computer system, which is analogous to the switching center introduced here.

Information Flow

In an office system, information can flow in several directions depending on the functionality of the system and the nature of the components involved. Five basic patterns will be discussed:

Terminal and terminal—bidirectional and simultaneous.
Terminal and center—bidirectional with delay.

Figure 5-4. With a switching center, only n (or five in this case) communication links are required.

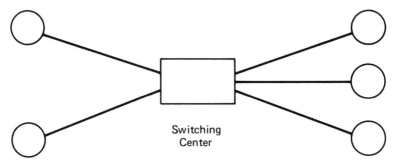

Switching
Center

Figure 5-5. For long distances, trunk lines are used for efficiency.

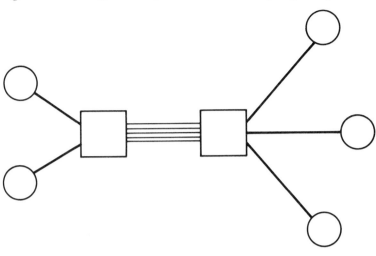

Terminal to terminal—unidirectional with delay.
Center to terminals—unidirectional.
Terminals to center—unidirectional.

Each pattern can be used to model a particular aspect of office systems design.

Terminal and terminal—bidirectional and simultaneous. This pattern of information flow, depicted in Figure 5-6(a), involves a synchronous connection between two end points, as in the case of a switched telephone line. Videoconferencing is one of the few office automation application that follows this pattern for information flow.

Terminal and center—bidirectional with delay. This pattern of information flow, depicted in Figure 5-6(b), involves a two-way connection. Delay represents computer processing and access to a data base. Characteristic of an on-line system, such as an airline reservation system, this pattern can be used to adequately model any office function where a computer responds to a keyboard input, as in a retrieval/query operation, a word processing interaction, or an

Figure 5-6. Patterns of information flow in a communications system. (Courtesy Petrocelli Books, Inc.)

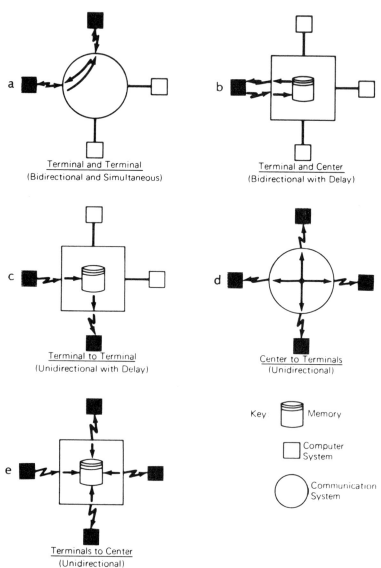

Terminal and Terminal
(Bidirectional and Simultaneous)

Terminal and Center
(Bidirectional with Delay)

Terminal to Terminal
(Unidirectional with Delay)

Center to Terminals
(Unidirectional)

Terminals to Center
(Unidirectional)

Key   Memory

Computer System

Communication System

entry of or a request for electronic mail. This is the prevalent form of information flow in distributed systems and in office systems; in fact, this pattern can be combined with other patterns to model a complex office environment.

Terminal to terminal—unidirectional with delay. This pattern, shown in Figure 5-6(c), represents electronic mail, conferencing, and other forms of office communications such as facsimile. This is the general class of message systems that includes telex, TWX, and computerized store-and-forward systems. In this pattern, information flows from source to destination by successively being stored, routed, and forwarded. The storage operation is an inherent part of message systems, and the delay may vary widely from fractions of seconds to hours.

Center to terminals—unidirectional. This pattern, shown in Figure 5-6(d), is characteristic of information dissemination systems such as radio and TV. Combined with a storage facility, this pattern suggests electronic mail that includes a distribution option for sending the source message to several parties.

Terminals to center—unidirectional. This pattern, depicted in Figure 5-6(e), represents a data collection operation as in some forms of computer conferencing. Normally, a terminals-to-center pattern is used in combination with other patterns.

Probably the most important point that should be made with regard to any of the patterns identified is that problems of information flow in an enterprise are not unique and that office automation systems can adequately serve almost any conceivable operational environment.

Computer Networks

A computer network is a collection of computers and terminal equipment connected by a communications system. The computers may include centralized computers and remote computers. The terminal equipment may or may not have intelligence and may include workstations designed for general or special purposes.

Figure 5-7. Centralized computer network structure.

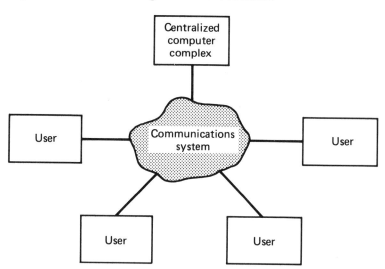

Computer Classification

A centralized computer that controls a significant portion of the processing is commonly known as the *host computer*. In actual practice, the host system can also be a set of processing units operating in tandem to do the work. Another name for host system is *mainframe*.

Computers connected to the host computer through telecommunications facilities are called *remote computers* for obvious reasons. As covered in a subsequent section, remote computers can be linked to a host system in a hierarchical manner or can exist as peers in an environment where the data processing is distributed.

Note that the terms are relative, since a host computer to a given workstation may be a remote computer to another.

Basic Network Structures

To a large extent, computer network structures are derived from two basic concepts, as suggested by Figures 5-7 and 5-8: cen-

Figure 5-8. Distributed computer network structure.

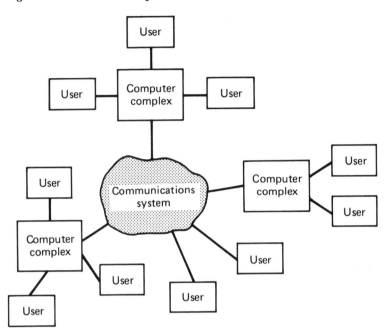

tralized and distributed computer network structures. The concepts appear to be rather similar on the surface, but in reality they are not, because the configurations differ, as do the modes of transmission.

A *centralized computer network structure* is characterized by a host computer, a communications system, and a set of terminal users who can interact with the computer system through local terminal equipment. Usually, the asynchronous mode of transmission is employed when the end points are data terminals.

A *distributed computer network structure* is characterized by two or more computer systems that are linked by the communications system. A user may interact with a local computer system for service and for accessing other nodes in the network, or the user may be connected directly to the communications system. When computers communicate with each other, the synchronous mode of transmission is used.

Public Network Service
For a relatively small network, especially one in an office environment for electronic mail or distributed office facilities, a local telephone connection, in-house communications system, or physical data link is more than satisfactory. When the communications system goes over long distances, however, the lines can be very cost-ineffective for low message volumes.

By effectively pooling resources, small users can use a communications system without having to finance the entire operation. A *value-added network* is a public network service that builds on existing telecommunications facilities and provides service for small users.

The result is that an enterprise need only connect computers and terminal equipment to the value-added network without having to develop a complete communications network structure.

Network Types

There are two basic types of networks: point-to-point and multipoint. They differ in how communications are handled by the computers involved. The selection of a network type is based on two considerations: line cost and line-handling technique.

Point-to-Point Connection
In a point-to-point connection two nodes—taken here to be computers or terminal equipment—are connected by a single communications link that exists for the duration of the session. When a salesperson at a remote location, for example, dials into a central computer for a quotation or a list of electronic-mail messages, a point-to-point connection is established through switched lines available from the telephone company.

In general, point-to-point connections can be developed through switched lines, private leased lines, or physical connections through local facilities. Portable terminal equipment, such as a computer in a briefcase, normally is designed to use switched lines. Private leased lines and physical connections are usually designed into communications systems with fixed end points.

Multipoint Connections

In a multipoint connection—also called a *multidrop line*—several workstations share a common communications link. During operation, however, only one workstation can transmit or receive messages at a time. This is a means of line sharing to reduce communications cost, and it can be implemented with private leased lines or physical connections. The control station in a network of this type is a computer that polls user stations on a periodic basis.

Network Configurations

Operational network configurations are derived from the basic network structures (centralized versus distributed) given earlier. A network can be developed with practically any conceivable network topology. The most widespread set of network topologies is covered here. The extreme variety of possible network configurations is suggested in Figure 5-9.

Star network. A star network is nothing more than a centralized network structure where communication between two remote nodes must pass through the host computer system. A slight variation to the star network is the *hierarchical network structure,* which is composed of several levels consisting of a central computer, remote computers, and workstations.

Distributed network. In a distributed network, each computer system can in principle be connected to every other computer system to achieve a high level of intersystem reliability. When each system is in fact connected to every other system through a communications link, the network is said to be fully connected. Line costs are high in a fully connected distributed network, so that partial connection is predominantly used in network configurations of this type.

Figure 5-9. Varieties of network configurations.

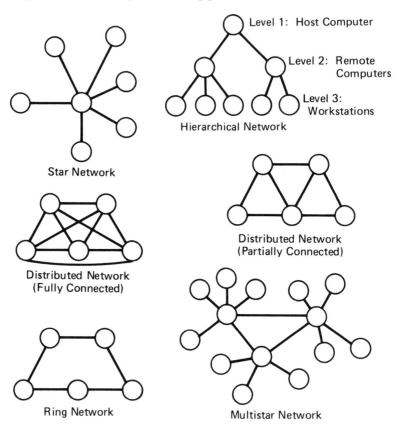

Ring network. A ring network is a special case of a distributed network in which each node is connected to two other and only two other nodes. A ring network is reasonably cost-effective but is susceptible to network failures.

Multistar network. A multistar network is a distributed network configuration in which each node is a local star network—hence the name "multistar." This configuration is commonly used with on-line systems for banking and insurance where the regional centers are connected by very fast communications lines.

Distributed Systems

Systems are usually distributed for two major reasons: to share resources and to give an increased level of local control. Shared resources are hardware facilities, such as a printer, and information. An increased level of local control permits close matching of the processing capability to the information needs of the organization.

The 80-20 Rule

The 80-20 rule is a means of emphasizing the manner in which information flows in most organizations. In the familiar organization structure, conveniently depicted as an organization chart, information is summarized and abstracted as it flows upward. Similarly, management control information is expanded and detailed as it flows downward. Thus the organization operates as an information processor of sorts, and most information passing through the processor is both generated locally and used locally. This is where the 80-20 rule comes from. It is estimated that 80 percent of the information used in a department is generated within the department and only 20 percent comes from outside. Conversely, 80 percent of the information processed by a department will be used only in that department and 20 percent will be transmitted outside that department.

The 80-20 rule is useful in determining where resources should be located for optimum performance.

Distributed Functions

In a distributed data processing or office system environment, three major application functions can be identified: information capture, information processing, and information management. *Information capture* refers to where information is entered, regardless of whether a point-of-sale (POS) terminal is being used for data processing or an office workstation is used for word processing. *Information processing* refers to the computation, reporting, text editing, or message handling in either of two major applications (data processing or office automation). *Information management* refers to whether the data base resides in a host computer, a remote computer, or an intelligent workstation.

When the term "distributed function" is used, it is important to recognize what the concept entails. There are two general ways in which functions can be distributed. In the first way, called *replicated,* the same function can be executed at more than one node in the network. Thus the processes of information capture, processing, and management are replicated throughout the network. In the second way, called *partitioned,* a function cannot be performed in a single node and the various parts are executed cooperatively in separate nodes.

Data Processing

Banking and retailing supply two examples of distributed data processing. In a banking system, intelligent teller workstations are connected to local "remote" computers for polling and for handling local data files and bookkeeping. There is a need for centralized record keeping in a banking environment, so that a host computer is needed to do the information management and end-of-day processing. The various components are commonly connected by communications facilities into a star or multistar network.

In a retailing system, point-of-sale (POS) devices are connected to a local "remote" computer for cash management and inventory control. There is a need for regional inventory management and centralized accounting control in a retailing environment, so that regional and host computers are needed to do auxiliary processing. The various components are frequently connected by communications facilities into a star network.

Office Systems

Word processing and electronic message systems supply two examples of distributed office systems. Other than stand-alone word processing machines, office automation conveniently lends itself to a distributed environment, and many office automation functions, such as electronic mail, would not be possible without computer networks.

In a *distributed word processing system,* two distinct possibilities are readily apparent: (1) shared resources, in which several stand-alone workstations share a common resource, such as a printer, and (2) concurrent support of several non-stand-alone workstations by a remote computer. In the first possibility, the stand-alone workstations constitute a semidistributed system in that the only network commonality is the shared printer. In the second possibility, the workstations clustered about a remote computer compose a simple star network; if the local star network is further linked to other networks, then perhaps the total configuration is a multistar system.

An *electronic message system* is by definition a distributed office system. As with the second word processing possibility, a configuration can be a star network, a multistar network, or one of the variations of a distributed configuration.

The subject of distributed office systems is further amplified in later chapters.

Summary

Data communications facilities provide the capability for transferring data between remote locations and are integrated with data processing and office automation in modern office systems. The age-old model of a communication system consisting of a message source, an encoder, a communications channel, a decoder, and a message destination still serves as a good basis for analysis. In the use of modern telephone systems, the modulator/demodulator serves as the encoder and decoder, and the telephone network is the communications channel.

There are two modes of transmission: asynchronous and synchronous. They apply to message transmission as well as to other office communications functions. Switching and information flow concepts provide a means of designing and analyzing network models, respectively.

One of the most important components in a computer network structure is the computer. Computers in a network can be classified as host computers (or mainframe) and remote computers. Two fundamental network structures serve as a basis for network development: centralized and distributed. Value-added networks provide a means for relatively small organizations to derive the benefits of networking.

Network connections can be point-to-point and multipoint, depending on whether lines are shared or not. Depending on the network structure and methods of connection, several network configurations are possible: star network, hierarchical network, distributed network, ring network, and multistar network.

The success of distributed systems is in part a result of the 80–20 rule, which states that 80 percent of the information needed in a department is generated in the department. Distributed functions are information capture, information processing, and information management. In general, these functions can be replicated or partitioned throughout the network. Two examples of distributed data processing systems relate to banking and retailing. Two examples of distributed office systems are associated with word processing and electronic message systems.

SUGGESTED READING

Bingham, J. E., and Davies, G. W. P., *Planning for Data Communications*, London: Macmillan, 1977.

Cypser, R. J., *Communications Architecture for Distributed Systems*, Reading, MA: Addison-Wesley, 1978.

Doll, D. R., *Data Communications: Facilities, Networks, and Systems Design*, New York: Wiley, 1978.

Katzan, H., *Introduction to Computers and Data Processing*, New York: D. Van Nostrand, 1979 (Chapter 22: "Computer Networks and Distributed Systems," pp. 454–469).

Katzan, H., *An Introduction to Distributed Data Processing*, Princeton, NJ: Petrocelli Books, 1979.

PART III
Topics in Office Automation

CHAPTER 6

Word Processing

Word processing is a necessary ingredient in a total office automation project. While the overall concept is reasonably straightforward, the inherent simplicity of word processing disguises many salient points that can make realistic expectations practically unachievable. Moreover, the number of hardware and software products in this category is quite large, so that checklists and guidelines are particularly useful. This chapter gives the key features in word processing that can aid in selecting an effective word processing system. It does not cover planning for word processing, which has been broken off and placed in a separate chapter on that subject.

Basic Technology

The basic technology of word processing involves the underlying computer equipment together with the manner by which the operator interacts with the system. Fundamental decisions must in general be made in this latter category before equipment selection.

Dedicated and General-Purpose Systems

When a computer is selected for a particular application and is used only for that application, it is known as a *dedicated system.* This concept can be applied to word processing in two ways: (1) the only software available to the user supports word processing or (2) the computer itself cannot be programmed by the user and must be used only for word processing.

Some stand-alone word processors are dedicated, but this is not a definitive characteristic. A dedicated word processor is neither good nor bad. Clearly, it is impossible to do electronic mail with a dedicated word processor, but it is possible to physically transport a diskette from one unit to another. For a given application domain, however, "physical communication" may be all that is needed.

Some vendors supply dedicated systems that have been packaged especially for word processing. Normally, a computer is paired with appropriate software and is sold and serviced as a single unit. Vendors in this category can be computer manufacturers, software houses, systems houses, and computer stores.

A *general-purpose system* is a computer system that can perform data processing, problem solving, and office automation functions in addition to word processing. In most cases, an enterprise will buy other software for these additional applications, and in some cases even write the programs themselves. General-purpose systems can be stand-alone or multiterminal word processors, but again, this is not a definitive characteristic.

Most general-purpose systems are acquired from computer manufacturers or computer stores. It is also possible to obtain a computer used as a general-purpose system from a software or systems house. In the latter case the outside vendor will also supply the diverse application programs.

Software is available from computer vendors and computer stores for major applications. In fact, the most difficult decision to make in some cases of word processing can be the choice of software.

Business Computers and Personal Computers

Software for word processing is available for business computers and also for personal computers. A brief visit to a local computer store will verify that a variety of word processing packages are available "off the shelf."

The primary difference between business computers and personal computers usually lies in raw processing power, memory, mass storage capability, and the range of input and output units.

There can also be a big price difference between business and personal computers. Personal computers are usually lower in price, but not always.

There was a time when good sales and service were available for business computers but not for personal computers. This may still be true in some cases; however, many computer stores currently offer service arrangements that are comparable to those of established computer manufacturers.

Another consideration is education and training. Many users feel that this area is of prime significance. While good training and reference manuals are important, setup, demonstrations, and hands-on training often spell the difference between user acceptance and rejection.

There has been a recent trend toward providing business applications software with word processing as well as providing word processing applications software with traditional business applications. Personal computers span both areas, because software currently exists for business computing, word processing, and personal computing. Some word processing systems can even be used as personal computers by loading an appropriate set of operating systems programs. While the use of word processing facilities for personal computing does not at this time constitute a major trend, it serves to illustrate that the various inherent technologies are reasonably common.

Keyboard/Display Technology

Existing keyboard/display technology for word processing can be divided into three categories: (1) electric typewriter systems, (2) thin-window display systems, and (3) video display systems.

An *electric typewriter system* features local storage and editing features and is, in some cases, connected to computer facilities through communications facilities.

A *thin-window system* consists of a single unit that contains a keyboard, a printer, and a single-line display unit. The display can be used to review a document by recalling one line at a time. Editing the storage facilities are also available.

By far the most popular keyboard/display technology in modern word processing systems incorporates a *video display*. When connected to a workstation with intelligence, the display is normally driven by a buffer in RAM. When the display is connected to a dumb workstation, the display control usually contains a buffer memory.

Stand-alone, multiterminal, and business systems use a full video display that can represent 24 or 25 lines of 80 characters. The video display technology is integrated into the design of the workstation and incorporates features that are "human-engineered" for user acceptance.

A word processing system running on a personal computer almost always uses a video display, which can be a black-and-white monitor, a color monitor, a TV set, or an ergonomically designed monitor. Some personal computers have display limitations of 40 characters per line, which impairs visual fidelity for word processing. Other systems include full 80-character line capability. Most 40-character systems can be enhanced to 80 characters with a special circuit board.

Configurations

The fact that word processing capability is available in the form of stand-alone and multiterminal systems naturally leads to a consideration of the characteristics of each of these configurations. To a large extent, this section reviews and summarizes the information presented earlier on this important topic. This section also uses terminology intended for the presentation of the subject matter but relevant to the field of office automation in general.

Terminology

In the preceding chapter, a distinction was made between host and remote computer systems. A host computer is called a "mainframe" because it is the main computer system in a data processing shop. Certainly people can use this terminology in any way they desire, but "mainframe" should be used to refer to a powerful processing facility with a wide range of peripheral equipment such as input/output units, special control devices, and so forth. A remote computer, on the other hand, is not as powerful in processing capability as a mainframe, but it is not a microcomputer either. It is clear that distinctions among the various classes of computers are necessary and should be made.

A *microprocessor* is a processing unit on one or more electronic chips. Most microprocessors are a couple of inches long and an inch or so wide. Some are smaller. A microprocessor packaged with ROM, RAM, and a means of attaching some devices such as disks and printers is called a *microcomputer* (or *personal computer*). Microcomputers generally have limited capability because of the speed of the processing unit, the size of RAM, and the input/output capability. *However, modern microcomputers do have sufficient capability to handle word processing successfully.* Microcomputers can be and are used for small-business data processing, word processing, personal computing, and other technologically based applications. Most microcomputer systems are single-user-oriented, meaning they can perform only one application function at a time.

The next step up from microcomputers leads to *minicomputers*. A minicomputer offers powerful processing capability with the option to integrate a wide range of input and output devices. If a proper operating system is available, a minicomputer can service several users concurrently by switching between them at high rates of speed. While minicomputers are small in size—about the size of a desk or possibly smaller, depending on the packaging—they are generally regarded as being considerably more versatile than microcomputers.

A *remote computer* (or *distributed processsor*, as it is sometimes called) is somewhere between a minicomputer and a mainframe. Remote computers customarily have the capacity for data processing, communications, and office automation, although software for office automation may not be available for some systems. Remote computers are usually desk-size (or a little larger) and can drive sophisticated reprographics equipment. Distributed-processing software can support several concurrent users and can handle a diverse set of application programs.

Figure 6-1. Stand-alone word processing system. (Courtesy International Business Machines Corporation.)

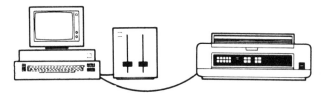

Stand-Alone Word Processing Systems

Stand-alone word processing systems are available on small minicomputers and microcomputers and incorporate a display unit, a keyboard, a processing unit and memory, a mass storage device, and a printer. Stand-alone systems may be connected to share a resource, such as a printer, in which case they are called *shared-resource systems.* Figure 6-1 depicts a stand-alone unit and Figure 6-2 a stand-alone shared-resource system sharing a printer. In most, if not all, systems where stand-alone systems are connected to share a component, the shared resource is a printer.

Multiterminal Word Processing Systems

Multiterminal word processing systems are available on host computers, remote computers, and minicomputers. The computing facility services several workstations, which may be intelligent, semi-intelligent, or dumb. Most word processing systems in the category represent either of two design philosophies: shared-logic systems or concurrent systems.

In a *shared-logic system,* the computer facility is designed to service two or more workstations by providing word processing functions. Mass storage may exist at the workstation or may be provided at the computer site. The term "shared logic" refers to the control and processing aspects of word processing, and in systems of this type, other computer applications are not executed at the same time as word processing.

In a *concurrent system,* the computer facility is used to handle other applications at the same time it is doing word processing.

Figure 6-2. Stand-alone units connected in a shared-resource system.
(*Courtesy International Business Machines Corporation.*)

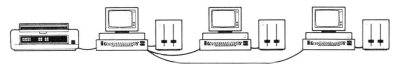

Moreover, these additional applications may include other office automation tasks in addition to word processing.

In both shared-logic and concurrent systems, documents stored at the computer facility can, in general, be shared among users as governed by installation guidelines. A multiterminal system is conceptualized in Figure 6-3.

Global Functions

Global functions in word processing are concerned with how the operator approaches the system and the facilities that are available for dealing with documents as a whole or in part. This section is not all-inclusive but covers most of the important considerations in word processing. New features are continually being developed, and it is almost impossible to keep up with them.

Storage Medium

The storage medium for the long-term storage of documents is of primary concern in word processing because it implicitly determines the size of documents that can be created and how many documents can be stored per volume of storage. (A diskette, for example, is regarded as a storage volume.) Documents that are stored in formatted form generally take more mass storage space than documents stored simply as a string of characters and editing commands. Stand-alone systems frequently use diskettes for storage, and total capacity may be of concern in systems design. Multiterminal systems use hard disks, which have a greater capacity for storage, so that the storage medium is normally less of a concern.

Figure 6-3. Conceptual view of a multiterminal system.

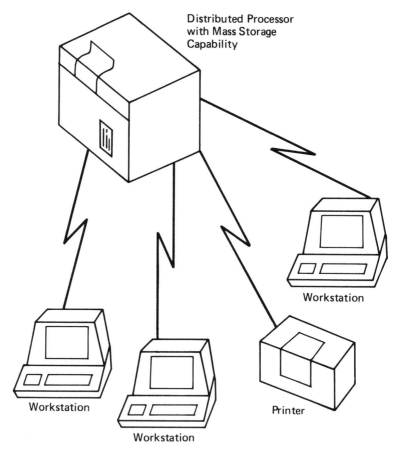

Distributed Processor
with Mass Storage
Capability

Workstation

Workstation

Workstation

Printer

Display Format

The subject of 40-character and 80-character displays was covered previously. Some word processing systems display a document in unformatted form as a file of characters. This method uses storage in an efficient manner but is lacking in the visual fidelity available with page-formatted displays. When documents are displayed as they are formatted, the image on the screen is the same as will be printed, so the operator can identify page breaks and other aspects of text arrangement. This is a major consideration in the selection of word processing systems. "What you see is what you get" is an important aspect of user acceptance.

File Processing

In word processing, file processing refers to the capability of storing segments of text as data files and later merging them into a document being created. Thus, frequently used paragraphs or sentences can be saved as files and conveniently added to letters or contracts.

Printer Overlap

A useful feature that is available with many word processing systems is the ability to complete one document and put it into a queue for printing while working on another document. That is, the operator can create a document at the keyboard while another is being printed. For years this capability was available only on expensive stand-alone and multiterminal systems. With recent advances in microcomputer software, however, this feature is currently available with many microcomputer systems.

Information Management

One of the primary objectives of a word processing system is not to require that a user have a computer background in order to be efficient and effective at assigned tasks. While it is practically impossible to completely eliminate the computer concept from word processing, certain aspects of a computer environment seem to sit on the fence between the two disciplines. A few areas of normal concern are:

Duplication of text files and diskettes.
Deletion of unused information from diskettes.
Listing of the directory of the contents of a diskette.
Giving the unused capacity of a diskette.

Software facilities for performing these tasks are normally provided as part of the word processing package.

Editing Functions

This section gives an overview of editing functions in word processing systems. As with the previous section, it is practically impossible to list all topics in this category; however, the major points are covered.

Editing functions are concerned with screen management and how text is edited. Clearly, there is a fine line between editing and formatting, which is covered in the following section. Some topics grouped under formatting with some word processors are considered to be editing here, and vice versa.

Screen Management

Two items are normally of concern as far as screen management is concerned: the text that is present on the screen and the position of the cursor. Text positioning is known as *scrolling,* and it determines the text that can be viewed by the operator. Two options are available in this regard: (1) the text can be viewed by page in either a forward or backward direction or (2) the text can be moved up or down by line in either a forward or backward direction.

Scrolling can be menu-driven or command-driven, with the following options: line up, line down, screen up, and screen down. Regardless of the text displayed, editing is keyed to the position of the cursor on the screen.

The *cursor* usually takes the form of a blinking underline, a blinking light-colored rectangle, or a similar positional indicator. Whenever an insertion or deletion of textual information is made, the position of the cursor on the screen and the word processing software establish the association between stored and displayed text.

Several cursor control options are normally available:

Character left.
Character right.
Word left.

Word right.
Line up.
Line down.

When an attempt is made to move the cursor past the left edge of the screen, it becomes immobile. When an attempt is made to move the cursor past the right boundary, it is moved to the left boundary but down one line. When the top or bottom boundary is reached, down scrolling and up scrolling operations of one line, respectively, are invoked. (See *vertical scrolling* in Glossary.)

When a word processing system permits documents with a width greater than screen width in characters to be created, stored, edited, and printed, the process of viewing a line of text in relation to other lines becomes a problem. This situation applies even more often to some small computer systems where the maximum screen width is only 40 characters. In systems of this type, horizontal scrolling is normally available through a "toggle" function key. When the key is toggled, you get the right half of the text window. When it is toggled again, the left half is shown.

The text window is a critical factor in user acceptance. A few scrolling operations are:

Move cursor to beginning of text file.
Move cursor to end of text file.
Scroll down a specified number of lines, such as 20 or 30.

The effectiveness of screen management operations is generally related to speed. Many operations consume so much processing and display time that they are simply not implemented.

Character and Word Operations
Character and word operations permit insertion and deletion of characters, words, lines, and sentences through the use of special keys or editing control functions. When an operation in this class is activated, it uses the current cursor position to determine where the insertion or deletion should be made.

Search and Replace

A search-and-replace operation is designed to locate a series of characters in a document. Once a segment of text is located, two options are available: (1) replace the text with another automatically, or (2) suspend the operation pending a decision from the user either to replace, continue search, or abort the operation. In the replacement mode, text is normally located and replaced for the entire document or only for a particular occurrence.

Block Move

A block-move operation permits a block of text to be bracketed and moved to another location in the document or duplicated in another location in the document. The traditional "cut and paste" operation familiar to most writers is done electronically through a block move.

A variation to a block-move operation permits a segment of text to be written to a file, from where it can be subsequently inserted in several documents.

Key Phrase

The key-phrase concept permits a phrase to be stored in (or held in) RAM and assigned a function key or some identifying characters. When the function key or identification is entered, the phrase is entered at the current cursor position.

This function can be emulated by entering identifying characters into the text during document creation and later performing a search and replace operation.

Formatting

Formatting functions may be performed when a document is created or when it is printed. Certain auxiliary operations, such as headings and footings, as well as page numbering, are almost always performed when a document is printed.

Specifications

Page specifications generally incorporate any aspect of document preparation that can be established for an entire document. Some of the most commonly used specifications are:

Line spacing.
Margins.
Lines per page.
Tab positions.
Headings and footings.
Page numbering.
Footnotes.
Superscripts and subscripts.

In addition, most comprehensive word processors provide the capability for right-hand and left-hand numbering corresponding to odd- and even-numbered pages, respectively. Multiple line headings and footings are not common but are available with some comprehensive word processing packages.

Text Positioning

As in publishing and precision typing, text can be positioned in a line on a page in any of several familiar forms:

Centered.
Left justified and ragged right.
Ragged left and right justified.
Left and right justified.

When left and right justification is invoked, spaces are normally placed between words to achieve the desired format. *Proportional spacing* provides for characters of differing widths, as are commonly available with executive typewriters, daisywheel printers, and a few matrix printers.

An advanced facility supported by some word processing packages is known as *microspacing*. With microspacing, microspaces are placed between letters to achieve justification, eliminating the need to place multiple spaces between words. When microspacing is combined with a letter-quality printer, the printed text approaches typeset material.

Wordwrap

Wordwrap refers to what happens when the text reaches the end of the line. With word processing, a user enters text without hitting the return key as would be the case with an electric typewriter. When a word extends beyond the right margin, it is automatically placed on the succeeding line by the word processing software. This feature increases typing speed and facilitates document creation.

Hot Zone

When a document is created using a ragged-right format, some word processors recognize what is known as the "hot zone." A hot zone is defined as a distance in characters from the right margin. When a word enters the hot zone, it is automatically placed on the next line.

Spelling

A standard feature on some word processing software that is available as an option on others is the capability to check the spelling of words in a document. A standard vocabulary is provided. Words peculiar to a particular application domain—such as medicine, law, or computer science—can be added to an auxiliary dictionary.

Automatic hyphenation is available on some advanced word processing systems, and the feature can normally be turned on and off. This feature, which requires a comprehensive dictionary and generally more processing time, supplements the justification feature. Thus hyphenation is performed automatically when the function is invoked.

Reformatting and Manual Hyphenation

Through the use of wordwrap and hot-zone facilities, the display can represent final copy in the proper operational environment. A series of insertion, deletion, and other editing moves can disrupt the final format. In this case, a paragraph must be reformatted as determined by the given specifications. With automatic hyphenation capability, a paragraph can be hyphenated as in straightforward document creation. Without automatic hyphenation, some word processing systems pause for each line with a potential hyphenation situation to allow it to be performed manually. When this occurs, the software normally permits a hyphen to be placed anywhere in the targeted word, providing justification as required.

Hard and Soft Characters

A *hard character* is a character explicitly inserted by the user; a *soft character* is one inserted by the editor for spacing, hyphenation, or line termination. When text is rearranged, as for example in reformatting, soft characters are obvious candidates for removal whereas hard characters are not. Thus, when the user enters a hard carriage return to end a paragraph, the paragraph ending remains invariant under all types of reformatting. Soft spaces are entered by the editor for justification and by the editor or user for hyphenation.

A related concept is that of a *required space*. In some legal contexts, for example, the phrase "July 4, 1983" must always be printed on one line. By inserting required spaces after "July" and after the comma, the user can ensure that the phrase will never be broken in reformatting. If it comes too near the end of a line, the entire phrase will be moved down to the next line.

Type Ahead

During document creation and some forms of editing, an expert typist can easily get ahead of the refresh rate of the screen. This is usually the case because new lines are added at the bottom of the screen and every other line must be moved up. Most word processing packages use text buffering, which guarantees that characters entered are saved in RAM even though they may not be physically evident on the screen. Textual overrun is not, in general, a problem, and the display unit quickly catches up with the user.

Printing

Printing refers to the process of "getting it down on paper." Even though a major objective of word processing may be to enhance an electronic document library for search and retrieval, conventional text creation is almost always a central function. Text-positioning functions can take place during formatting or during printing, depending on the word processing package. Proportional spacing and microspacing, covered earlier under formatting, may not be in evidence on the screen but may be visible on the typed page. Other "visual" features, such as superscripts and subscripts, may not show on the screen in any particular word processing system, and the hard copy may be the only means of guaranteeing a successful operation.

This topic reflects a general problem in word processing, since the more features are left until printing, the greater the chances of having to print a document more than once. It is definitely preferable to have as many specifications reflected on the screen as is physically possible.

Most of this section reflects rather obvious printing features that are normally available with word processing systems.

Physical Considerations

In order to achieve professional-quality printing, most printers, regardless of the technology, permit the following options in text formation: double strike, boldface, and overstrike. *Double strike* refers to printing the same image twice, thereby yielding a darker copy. With *boldface*, the image is displaced slightly, thus creating a wider character line and the visual impression of a thicker character. *Overstrike* permits a line to be overprinted with additional characters to synthesize special symbols and typographic impressions.

In each of these cases, a line is printed more than once without advancing the carrier. Software is necessary to command the printer to perform the required action. Overprinting is time-consuming with unidirectional printers but is reasonably convenient with bidirectional printers since the overstrike is performed on the return action.

Paper

Conventional continuous-form fan-fold paper is standard with most printers, and some even permit roll paper through the use of special attachments.

An option permitting single sheets is convenient for generating business reports, letters, and promotional material. The software halts after each page to allow the operator to insert another sheet. Thus, letterheads and other preprinted forms can be used. One- and two-form single-sheet feeders are also available for some printers—usually of the more expensive daisywheel variety—eliminating the need to enter single sheets. The choice of a primary form and a secondary form is made under software control.

Special forms can also be printed on or "waxed on" continuous paper so that single-sheet capability is available with continuous-form printers.

Copies

The ability to print multiple copies is a convenience factor for long documents, but a practical necessity for short documents, such as form letters. An operational necessity for report production is the ability to print a single page without affecting other pages. Clearly, this concept can be applied to several pages, since substantial revisions can ripple through several pages. Then, only the modified pages need be reprinted. This feature is a definite plus for word processors that display and store text in paginated form.

Advanced Facilities

Advanced features in word processing are based in most cases on host-level, multiterminal, or sophisticated minicomputer facilities. Only a few highly capable word processing systems offer all the advanced facilities described here, but many systems have at least one of them.

Columnar and Arithmetic Processing

The problem domain for columnar and arithmetic processing is a report containing columns of numbers. Through columnar features, numbers can be neatly arranged in columns aligned by decimal point or the right or left edge of a number. Moreover, columns may be deleted, exchanged, moved, or copied in the report. Arithmetic facilities permit arithmetic operations on columns of numbers and include placing the result in a specified position. Other arithmetic facilities permit arithmetic operations on values stored in the body of the text. In most systems, values are identified by cursor position together with arithmetic operators entered at the keyboard.

Record Processing

In certain documents, each line represents a unique entry, such as a person, project, or item of inventory. As in data processing, a line of this type is regarded as a record and the entire document as a file. A typical example of a file of this type is a telephone directory where each entry (or record) represents a person.

Record-processing facilities in word processing provide the capability of:

Sorting records.
Deleting records.
Adding records.
Merging records.
Modifying records.
Converting straight text to records.
Converting records to straight text.

As word processing facilities are integrated with data processing in a concurrent system, the two capabilities tend to merge to a combined information management facility.

Text Management

A common capability in advanced word processing systems is that of rearranging blocks of text—not only columns and lines, but also any information depicted on the screen. One method of doing this is to highlight the text to be rearranged and, through the use of the cursor and function keys, place it in another position.

Closely related to rearrangement is the ability to include or not include "blind text" in a letter or document. *Blind text* is information that should be concealed from certain recipients and revealed to others. Blind text is encapsulated in control characters so the software can recognize its special significance during printing and distribution.

Document Storage

Three types of storage are recognized in word processing: working storage, permanent storage, and archival storage. *Working storage* is that part of RAM and mass storage (that is, disk or diskette) used to hold a document when it is being worked on. *Permanent storage* is mass storage used to hold a document when it is not being worked on. *Archival storage* is where documents are placed when the permanent storage facility is saturated and the documents in question are rarely used. Archival storage for a host system, for example, could be magnetic tape or diskette. Documents can be moved freely among the various kinds of storage under user control.

Communications

The communications feature in word processing permits a document to be created at one location and retrieved via asynchronous or synchronous transmission from another word processing facility. A desirable feature in this environment is to have the format and content of a document be independent of the specific equipment on either end of the communications link. This is not an electronic storage problem but rather a problem of establishing a standard format for exchangeable documents. The result is achieved through the use of appropriately placed control codes in the document itself.

A document exchangeability standard is not limited to the hardware and software of a single vendor, although such things traditionally start with one forward-looking organization. Once established, document interchange facilities can serve to allow connection by communications links of the word processing and other office automation offerings of several vendors.

Summary

While word processing is a practical necessity in an office automation system, its inherent simplicity disguises many subtle points that sometimes obstruct satisfactory results. Word processing can be implemented on dedicated and general-purpose computers in stand-alone, shared-resource, multiterminal, and concurrent systems. There is a trend to integrate word processing with data processing and vice versa.

The key points overall in the selection and use of a word processing system include the following: storage medium, display format, file processing capability, printer overlap, and computer facilities for information management. A word processing system's capabilities for editing, formatting, and printing (as well as some advanced features) determine the functions for which it is best suited.

Editing features normally of interest are the methods of screen management, such as scrolling and cursor control, character and word operations, such as insertion and deletion, search-and-replace facilities, block moves, and the use of key phrases. Formatting features concern page specifications, such as spacing and margins, text positioning, wordwrap and hot-zone facilities, spelling and hyphenation, and the use of hard and soft characters. Printing features involve operational characteristics, such as boldface and overstriking, paper, and copies. Advanced features include columnar and arithmetic processing, record processing, text management, document storage, and communications facilities.

SUGGESTED READING

Boyer, R. D., *Computer Word Processing: Do You Want It?* Indianapolis, IN: Que Corporation, 1981.

Zarrella, J., *Word Processing and Text Editing,* Suisun City, CA: Microcomputer Applications, 1980.

CHAPTER 7

Electronic Mail

On the average, a knowledge worker spends approximately 50 percent of his or her time in meetings, including time spent on the telephone and other forms of communication. Many transactions require face-to-face meetings, but a high percentage of everyday business activities within the communication domain can take place in the asynchronous mode, which is more efficient and more convenient. Only a small fraction of the time spent in so-called communication is productive, because of extraneous conversation and other nonproductive behavior. Another consideration that is often neglected is that communication between some people is cumbersome and the parties involved would prefer a more formal means of communication. This sets the stage for message systems. There are other pressing reasons for electronic mail, ranging from "it's the only way" to "it saves time when we're in a hurry." This chapter surveys electronic mail and covers many of the reasons that modern organizations and the computer industry view electronic mail as a major growth area.

Dimensions of Electronic Mail

Electronic mail systems are characterized by the fact that they provide a fast, reliable, and distance-independent means of mov-

104

ing information from one place to another. Electronic mail represents nonsimultaneous communication—also referred to as asynchronous communication—but, unlike some other forms of communication, leaves a permanent record of the transaction.

While computer networks provide electronic mail in the most general sense of the word, the term is normally restricted in office automation to refer to one or more of four areas: (1) facsimile, (2) document distribution through communicating word processors, (3) message switching, and (4) electronic message systems. Message switching is a special case of computer-based systems and is not discussed further here. Facsimile, document distribution, and electronic message systems are strictly within the domain of office systems.

Facsimile

The name "facsimile" is commonly associated with newspaper and law enforcement, where wirephotos are sent over distances at relatively low speeds—that is, low compared with modern telecommunications facilities, but high relative to the postal service. In the past, slowness and poor photo copy were traditionally associated with facsimile transmission; however, speed and quality are greatly improved in modern systems.

Basic Concept

Conceptually, a facsimile machine can be viewed as a copier that converts an image by photoelectric means to an electronic signal, which is used to send the original document to a remote location, where the image is regenerated. The version received is regarded as a "facsimile" of the sending copy.

The key advantage of facsimile is that the contents of any page can be transmitted—text, drawings, charts, photos, or hand-written material. No keyboard data entry is required, and the only other physical operation is to feed the source material into the machine.

Technology

Facsimile is a relatively old process; it was developed in the 1800s and continually refined since then. The modern process of doing a facsimile transmission consists of the following steps:

1. Scanning the source document and converting it to a representative set of electrical signals.
2. Modulating the signals for transmission over telephone lines.
3. Demodulating the signals on the receiving end of the communications channel.
4. Converting the electrical signals, representing the image, to hard-copy form.

Step 1 is performed by a document scanner called the transmitter, and step 4 is performed by a recorder or printer called a receiver. Steps 2 and 3 are performed by the communications facility and refer to modulation and demodulation, respectively. Modems are normally included in the facsimile equipment.

Facsimile devices come in three varieties: stand-alone transmitter; stand-alone receiver; and combined transmitter and receiver, called a *transceiver*. Several methods are used for both scanning and printing, and a discussion of that technology is beyond the scope of this book. It is convenient to conceptualize a "fax" machine as a supercopier with data communications facilities somewhere between the beginning and the end of the process.

Electrical Signals

Two methods of scanning an image and producing an electrical signal are generally used: analog and digital. With an *analog signal*, electrical signals represent a complete document—the white part as well as the dark part. The image is divided into picture elements, and a signal exists for each element, resulting in good resolution but slow speed.

With a *digital signal*, a document is scanned, but binary information (zeros and ones) is produced only for actual image content. Through a coding scheme, an image is compressed in the horizontal direction (between black and white areas across a line) and vertically (between lines) as well. Because of the compressed form of an image, transmission times for digital facsimile are lower than for analog facsimile.

Classes of Service

Facsimile services are grouped on the basis of transmission speeds and the method of modulation. Three groups are recognized internationally:

- Group 1: transmission speed of either four or six minutes per page, using frequency modulation for analog transmission.
- Group 2: transmission speed of two or three minutes per page, using amplitude modulation for analog transmission.
- Group 3: transmission speed of one minute or less per page for digital transmission, using bandwidth compression.

Facsimile equipment is easy to install, and within each class optional capabilities such as automatic dial, automatic answer, and automatic disconnect are available.

Computer Service

The use of facsimile with computer-based electronic mail is not widespread, and most facsimile service is regarded as a "noncomputer" office function. However, at least one office systems vendor offers an image transfer system employing a technology analogous to facsimile that permits images to be digitized, stored, transmitted, restored, and regenerated using computer facilities. Many office automation specialists are looking to the future when text, image, and voice data can be combined in a single document.

Document Distribution

Electronic document distribution using communicating word processors provides the ability to transmit a document to any network user. A distinction is made here between document distribution and electronic message systems that incorporate an expanded function for electronic communication. This subject is discussed further in Chapter 8, "Systems Concepts."

Functionality

With communicating word processors in an environment augmented by a computer network, documents can be exchanged between users in three rather obvious ways: (1) by physically moving the document to the destination, (2) by storing the document in a central repository, such as a distributed or host computer, and (3) by storing the document in a central repository and also physically moving the document to the destination.

In a local site with intelligence and mass storage capability, a local copy of the document can be maintained. However, in some systems the local terminal has neither intelligence (other than screen-editing facilities) nor storage capability, and the source document is stored in a distributed computer.

While document distribution facilities differ widely, the heart of most systems is a mail log maintained by an "electronic mail software package." When a user signs on to the mail system, a log is presented that gives the following kinds of information for current electronic mail:

> Item number.
> Date.
> Author.
> Brief description of the item.
> Current disposition.

The "current disposition" normally indicates whether a document is incoming or outgoing and its state of processing. For incoming electronic mail, priority and confirmation notices are usually supplied. For outgoing electronic mail, responses and confirmation of delivery—that is, return receipt—are presented on the screen.

Commonly associated with electronic document distribution is the capability for sending a document to a distribution list, to redistribute a document, and to append a reply to the original document and return it. The latter case suggests an inherent characteristic of electronic mail. It is commonplace to send and also receive abbreviated messages in an electronic mail environment. Thus the knowledge worker need not be a typist. However, when abbreviated messages are passing back and forth, it is sometimes necessary to be able to associate a reply with the original message.

This concept also applies to the redistribution of documents. A familiar practice is to attach a "buck slip" to a document and pass it on to the appropriate party. Usually the buck slip contains comments or suggestions on how to proceed.

Schedule and Agenda

A useful feature associated with some electronic mail systems is a schedule/agenda facility to aid in office management. In some systems, the schedule/agenda information has to be called up from a distribution computer. In others, the software is invoked when the user signs on for electronic mail.

The schedule/agenda is normally managed on a daily basis with a look-ahead feature for a specified period. Thus, by making a proper request, a list of scheduled activities can be generated for a specified period. Through the use of software in this area, the executive can:

- Block out periods of reserved time.
- Generate scheduled activities for a given period along with miscellaneous notes whenever appropriate.
- Issue reminders.
- Establish priority items and restrict a scheduling period to specified people.
- Store and retrieve special information.
- Execute desk calculator functions.

While schedule and agenda functions are not explicitly dependent on an electronic mail system, they are naturally compatible. It is very convenient to sign on to the system and have a schedule and also the day's mail presented at the same time.

Figure 7-1. Components of a simplified computer-based message system.

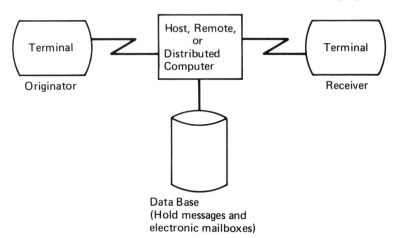

Electronic Message Systems

Electronic message systems have had an evolutionary development through the following technologies: telegraph, Teletype, telex/TWX, and finally computer-based message systems (CBMS). The components in a simplified CBMS are depicted in Figure 7-1; the connection between the terminal and the message processor can be communications lines or some form of in-house data link.

Information Float

One of the main advantages of computer-based message systems is that they take the information float out of communications. Using telex as an example, the following steps in the communication process might be experienced:

• Sender constructs message.
• Message is given to the word processing center for processing: one to two hours.

• Word processing center sends message to telex room for transcription: one to two hours.

• Message is sent and received: one minute.

• Telex room of receiver gets message for transcription and routing to recipient: one to two hours.

Thus a message that takes one minute to send may take up to four hours for processing; this is *information float.*

Cutting across time zones, evenings, weekends, and other informational bottlenecks, CBMSs help eliminate information float and distribute information, not more paper.

Office Productivity

The objective of computer-based message systems is to improve office productivity, *not* to produce documents for outside distribution. Thus it has a potentially different problem domain from document distribution.

Time savings for knowledge workers range from 30 minutes per day to two hours per day as a result of using CBMSs. Some of the benefits experienced are reduced wasted time, reduced interruptions, and increased quality of communication. The sum total is that more work gets done in less time and information is received without interruption. Since office costs are rising at a substantial rate, this increased productivity can easily be used to expand management's traditional span of control.

CBMS Functions

A computer-based message system is designed to handle text, as in word processing. It exists in three principal forms: with no word processing capability, with a built-in word processor, or with a link to word processing facilities. At the least, a CBMS contains a screen text editor that is used for constructing messages.

The functions performed by a CBMS allow messages to be created, filed, transmitted, and retrieved. However, a CBMS is more than the simple sum of its parts, because it provides a form of structured communication through generalized documents and forms, distribution lists, replying, and search facilities.

Generalized documents and forms is a concept where by the form of a document is predefined to the CBMS. When it is desired to construct a message of a given type, the user makes a selection from a menu and the desired structure is placed on the display screen. The user need only supply variable information and is prompted throughout the entire process of synthesizing a message. Typical classes of control information are: form identification, recipient, sender, subject, key words and phrases (specified separately as in the subject line of a memo), date, reply deadline, return receipt, distribution list, and then the body of the message. Subsequent search operations can be made on any of these fields. To search the body of the message, the computer uses a process called contextual scanning, which picks up key words and phrases and the body of the document.

User Interface

A computer-based message system is essentially an electronic office with facilities for message creation, filing, transmittal, and retrieval. Moreover, it is possible to be more explicit on the operational aspects of the user interface and regard the CBMS as an electronic desk, comprising the following set of capabilities:

In box.
Out box.
Current messages.
Electronic wastebasket.
Previous message.
File drawer.
Forms file.

Thus a user can initiate or suspend electronic mail activity and conceptualize the electronic desk as an ordinary desk. When work is resumed, the set of current messages—that is, those being worked on—is still active. However, between sign-off and the next sign-on, several changes would probably have taken place. Messages in the out box are transmitted. The electronic wastebasket is

Figure 7-2. Practical implementation of a CBMS.

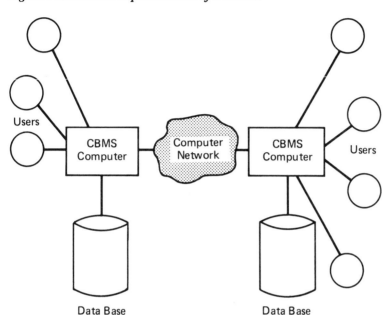

emptied. Incoming messages go into the in box. Whenever a specific form is needed, it is referenced from storage. Lastly, filing facilities are available for the permanent storage of messages.

Forwarding, reply, "buck slip" processing, archiving, and remote inquiry are additional functions that are part of the user interface.

Implementation

Figure 7-2 depicts a computer-based message system implemented through a computer network. This is the most practical approach, since a CBMS can be "piggybacked" onto an enterprise's data processing network. Thus a CBMS is a way of averaging the investment in a computer network.

The success of computer-based message systems lies in the integration of data processing, word processing, and electronic mail. The concept will most definitely include "voice mail" in the future, but the dimensions of the task of implementing a message system will remain about the same. A few organizations will implement stand-alone CBMSs, as well as stand-alone voice distribution systems. Specialists agree that both concepts can more than pay for themselves in the increased productivity of knowledge workers. Selling the concepts to management, however, requires a degree of strategic planning that is uncommon to the field of computers and data processing.

Summary

The average knowledge worker spends approximately 50 percent of the time in meetings, including telephone time and other forms of nonproductive communication. Many transactions require face-to-face contact, but a high percentage of everyday business activities within the communication domain can take place in the asynchronous mode, which is more efficient and more convenient.

Electronic mail systems provide fast, reliable, and distance-independent means of moving information from one place to another. Electronic mail is asynchronous but, unlike some other forms of communication, leaves a permanent record. In office automation, electronic mail refers to one or more of four areas: facsimile, document distribution, message switching, and electronic message systems.

Facsimile can be viewed as a supercopier with communications facilities. Two versions are recognized: analog and digital. Analog technology is mature, whereas digital methods are more recent in origin. Digital facsimile is more efficient and lends itself to a computer interface. The key advantage of facsimile is that the contents of any page—text, graphics, or handwritten material—can be transmitted. Most facsimile machines are installed as stand-alone equipment.

Document distribution commonly involves communicating word processors, augmented by a computer network. Document distribution may involve local or remote facilities—based on the nature of the local workstation and its capability for storing documents. Several systems options are available for moving documents between locations; however, the communications methodology is more a function of the total systems design than a matter of electronic mail operations. Features associated with document distribution systems are the comprehensive use of a mail log for document management and a schedule/agenda facility.

Electronic message systems are an evolutionary development of message-based systems such as Teletype and telex. Computer-based message systems (CBMSs) can be implemented as stand-alone systems or be integrated with data processing, word processing, and communications into total office environment. CBMSs reduce "information float" and improve the productivity of knowledge workers. While the functions of document distribution systems and CBMSs are similar, CBMSs are usually not intended for the production of comments for outside distribution. Through generalized document and forms control, a CBMS can be user-friendly and applicable for use by managers and executives. A CBMS may be combined with word processing or may contain a simple screen editor. One way of viewing a CBMS is that it is an electronic desk with an in box, an out box, an electronic wastebasket, forms, files, and a repository of "current" work. In most cases, a CBMS is a natural means of doing traditional office work.

SUGGESTED READING

Galitz, W. O., *Human Factors in Office Automation,* Atlanta, GA: Life Office Management Association, 1980.

Uhlig, R. P., D. J. Farber, and J. H. Bair, *The Office of the Future,* New York: North Holland, 1979.

CHAPTER 8

Systems Concepts

The systems approach is a way of thinking about systems that emphasizes the objectives of the system as a whole rather than its components. When dealing with information systems, it is important to recognize that it is the interaction of people, information, and the organization that is important. The interrelationship between components can take many different forms; in fact, this "interrelationship" is in itself a dynamic system that is constantly being restructured as events occur in the world of the enterprise. This chapter is concerned with the interrelationships that tie a system together. In short, we will cover three major topics: office systems design, packet switching, and local communications.

Systems Structure

The basic idea here is to look for as much precision in viewing the structure of office systems as is possible. The approach is qualitative, and the method is intended to demonstrate several dimensions:

Congruence—the most appropriate systems design concept is used at the proper time.

Integrity—the system provides a complete solution to the specifications.

Auditability—the system can be demonstrated to perform its intended function.

Controllability—the operation of the system can be effectively controlled by management.

Office systems are observable, capable of being described, and amenable to experimental manipulation to verify their congruence, integrity, auditability, and controllability.

Congruence. The dimension of congruence is related to the logic of systems analysis and the degree to which it reflects the relative capital–labor ratio. When people are replaced by machines, an enterprise becomes less tractable, but at the same time more deterministic. Thus the internal processes of an organization are greatly simplified and people are freed for more creative work.

Integrity. The dimension of integrity is related to the degree to which a system performs according to its specifications. In an office system, integrity refers to how a system will operate under normal conditions and also how it will operate when some components are not working or not used. It is an important element of planning to be able to specify, for example, how an office will operate in the event of computer systems failure. One way of measuring integrity is to test the *predictability* of a system.

Auditability. The dimension of auditability refers to the demonstrability of a system to or by individuals independent of the system. To meet reasonable tests of auditability, a system must be composed of auditable subsystems and the communication between subsystems must take place only by preestablished interfaces. Transactions must be recordable (called *journaling*) so that a reference to the external environment exists. A test of *accountability* refers to the ability to isolate the component that performs a particular event, and a test of *visibility* indicates that deviations from established patterns are readily observable by management.

Controllability. This dimension concerns the degree to which management can control the operation and also the use of a system. Two tests of controllability are granularity and specificity. The test of *granularity* is that a subsystem is designed so that its operation constitutes an acceptable risk to management. The test of *specificity* is the fact that the result of passing a resource from one subsystem to another is predictable.

To sum up, the functions performed by an office system must be describable, and its response to input conditions should be predictable.

Segmented and Differentiated Systems

A *segmented system* is one in which the components look alike structurally and also perform similar functions. A fleet of salespersons performing an identical function by selling the same product line is an example of a segmented system. A segmented system can have a flat structure or a hierarchical structure. A flat structure represents a few layers, each comprising many components. A hierarchical structure has a deep structure (many layers) with only a few components at each level. In both forms of a segmented system, all the components at the same level perform the same function.

A *differentiated system* is one in which different components perform different functions representing a degree of specialization. Collectively, a set of components performs an operational function, with a division of labor existing between the subsystems. A management or research team is a differentiated system because each team member contributes something different.

Office systems are both segmented and differentiated, depending upon their functionality and the method of interconnection.

Office Systems Design

Office systems can be viewed as interlocking organizational networks of activity networks. Thus the *organizational network* is the interoffice system connected by a telecommunications facility. The *activity network,* on the other hand, is an intraoffice system that can tie together the various office system components. The components in an activity network are linked by different kinds of connections, including a standard cable interface, a local network, and a tele-

Figure 8-1. Conceptual view of an activity network.

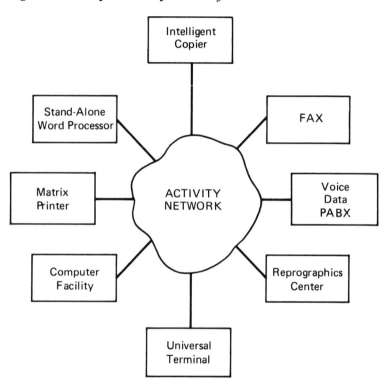

communications facility. Figures 8-1 and 8-2 suggest activity and organizational networks, respectively.

Methods of Interconnection

Chapter 5 on communications presents an overview of switching. The problem is essentially one of economics. When two communicating word processors are connected, for example, a single data link exists. When more than two word processors are connected, then either a ring structure or a loop structure can be established, provided that the workstations are in the same house.

Figure 8-2. Conceptual view of an organizational network.

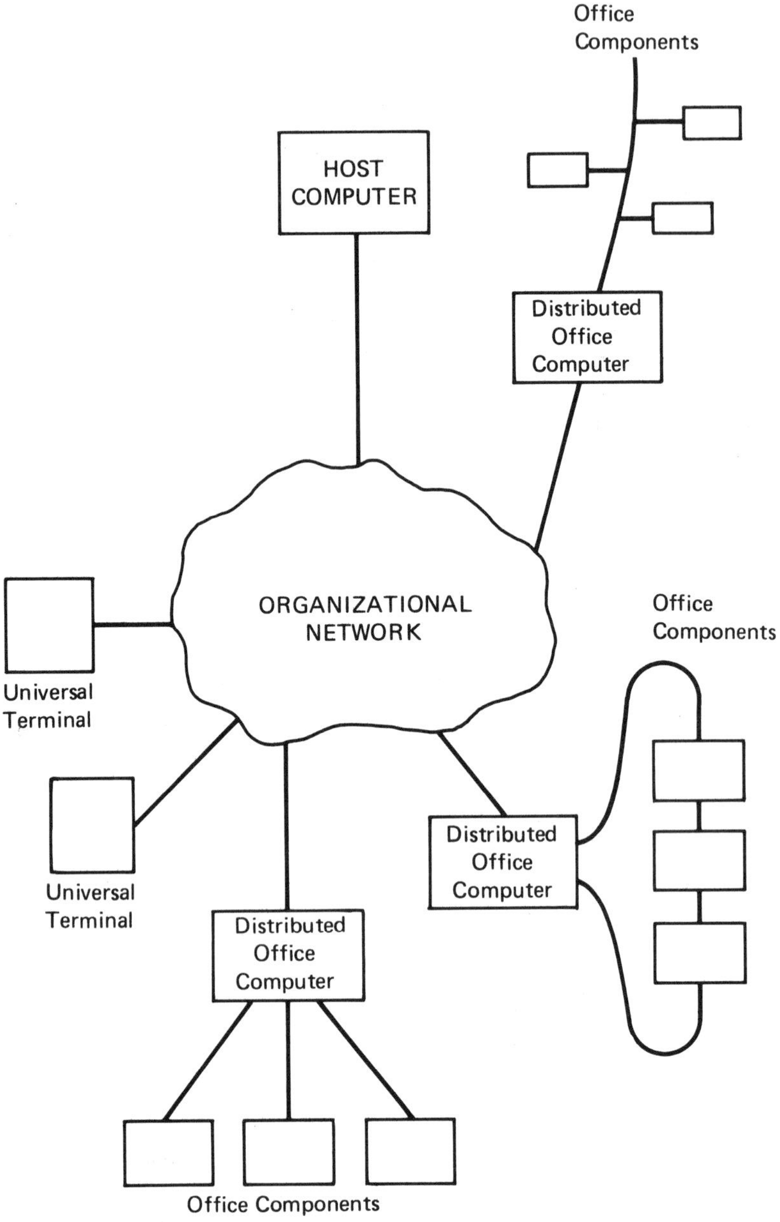

Figure 8-3. A distributed office computer serves as a switching center for communicating office workstations.

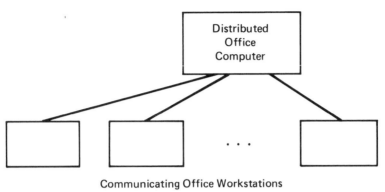

Communicating Office Workstations

Actually, multidropped data terminal equipment using telecommunications links is technically feasible, but the extent of its usage in office automation is not known. In most cases, communicating word processors in different houses will in fact require separate communication links.

Therefore, it seems reasonable to establish a switching center which, in fact, is a distributed office computer. In this case, two options exist:

1. The workstation is not a stand-alone system. The distributed office system drives the word processing, electronic mail, and other functions and also performs the communication between stations.
2. The workstation is a stand-alone system. The distributed office system handles communications, office electronic filing, and other support functions.

Figure 8-3 depicts both cases. A major advantage of this approach, in addition to its economy, is the fact that the distributed office computer can serve as a repository of information—electronic mail, documents, schedules, and so forth—as well as a source of processing functions. Figure 8-4 depicts an analogous configura-

Figure 8-4. A host computer can serve as a switching center for distributed office computers.

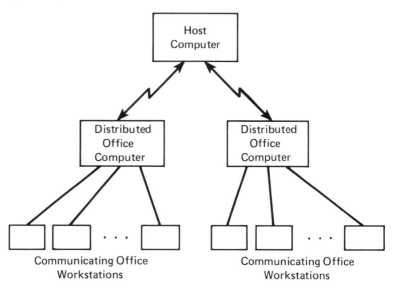

Communicating Office   Communicating Office
Workstations   Workstations

tion incorporating a host computer; the same advantages apply.
One method that is sometimes used to free a host of its commu-
nications load is to use a minicomputer as a *front-end communications computer*, as suggested in Figure 8-5. A front-end computer configured in this manner is used to implement communication between
distributed office computers without actually going through the
host computer itself. While the central-repository feature is usually
not available in this configuration, the economy of a switching center can still be obtained. Clearly, the host in this case can in principle also be used as a central repository of information and processing functions, as in previous cases, and it is the system
designer's responsibility to configure a viable system for an enterprise considering the dimensions of congruence, integrity, auditability, and controllability.

A final interconnection is given in Figure 8-6, which suggests a
feasible mode of communication between workstations connected
to different hosts. As before, the front-end computers are used to

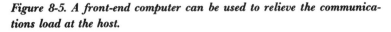

Figure 8-5. A front-end computer can be used to relieve the communications load at the host.

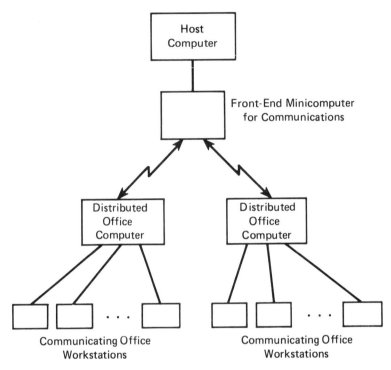

switch messages, and the messages can be stored in any of the distributed office computers or in either host computer.

The remainder of this section briefly reviews five major configurations: word processing, text/data integration, advanced text processing, information storage and retrieval, and electronic mail. The emphasis is on office automation, but as is clearly evident, it is practically impossible to separate form from function in communications-based computer systems.

Word Processing Configuration

A word processing configuration is a segmented system, characterized by the following components (see Figure 8-7):

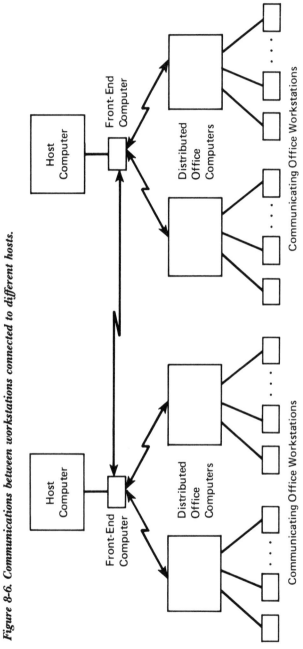

Figure 8-6. Communications between workstations connected to different hosts.

Figure 8-7. Word processing configuration.

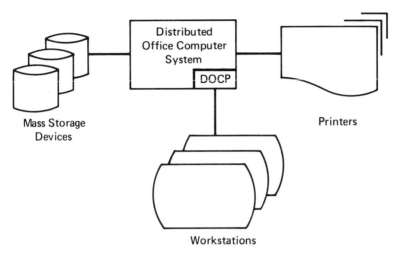

Key: DOCP = Distributed Office Control Program

A distributed computer system.
Mass storage capability.
One or more printers.
One or more workstations.

This is a modest configuration capable of serving a word processing environment comprising ten to 30 workstations and several printers, and providing the following kinds of functions: basic text processing, records processing, document distribution, and advanced text processing (spelling, automatic hyphenation, columnar and text arithmetic).

The office systems software is referred to here as DOCP—distributed office control program. It is designed to span the functions of a word processing configuration.

Text/Data Integration Configuration

A text/data integration configuration contains the same components as a word processing configuration, except that the hard-

Figure 8-8. Text/data integration configuration.

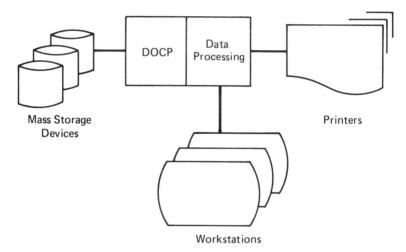

Key: DOCP = Distributed Office Control Program

ware system may be more powerful: more RAM, more mass storage, and possibly faster printers.

In this configuration, data processing runs concurrently with office automation software and allows access to text by data processing programs and access to data by word processing software. Therefore, pattern letters and mass mail are possible, as well as other office systems functions. This configuration is suggested by Figure 8-8. It is also a segmented system.

Advanced Text Processing Configuration

An advanced text processing configuration provides the computing power of a host computer, to be applied to an office environment when needed. Thus local text processing facilities can be augmented by host support in a differentiated system structure. This configuration is conceptualized in Figure 8-9. A differentiated system for advanced text processing can provide advanced formatting capability (such as multiple columns, footnote placement,

Figure 8-9. Advanced text processing configuration.

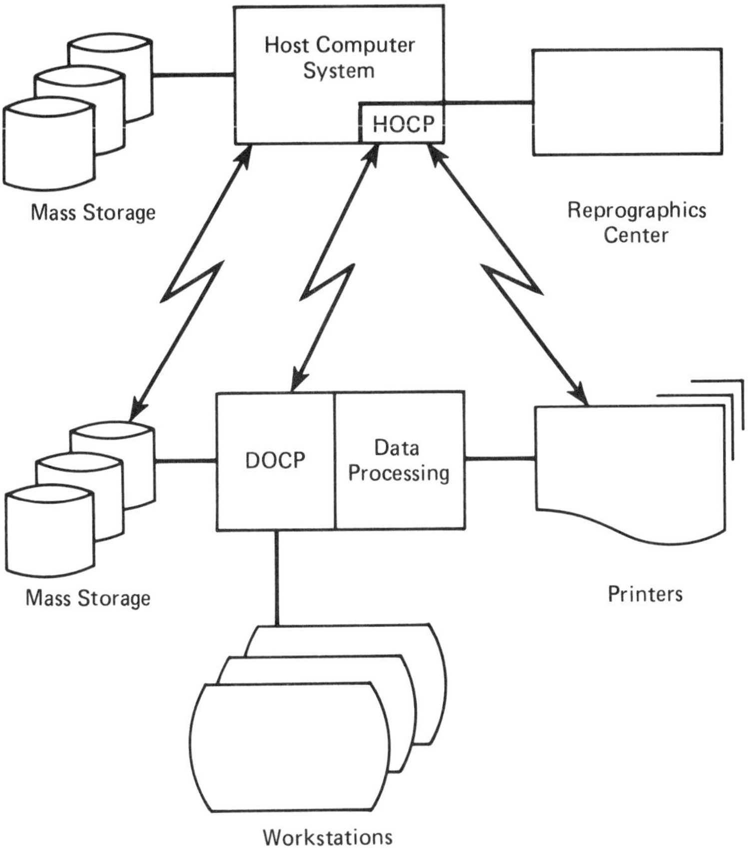

Key: HOCP = Host Office Control Program
     DOCP = Distributed Office Control Program

table of contents, indexing, and automatic spelling verification and hyphenation) and high-quality reprographics (such as multiple-copy capability using an advanced printer/copier, document distribution, multiple-sheet feeding and paper selection, font and pitch changing, line spacing, collating, and condensed printing).

Information Storage and Retrieval Configuration

An information storage and retrieval configuration (Figure 8-10) is a further refinement of the differentiated text processing configuration. In this case, the system is enhanced with a data base for electronic distribution and filing. The electronic filing system located in the host can perform the following operational functions in support of distributed computer workstations: index and contextual search operations, filing, automatic indexing, data base input and output formatting, and interface to the reprographics center for extended host distribution capability. As with previous configurations, information storage and retrieval operations are performed by software elements configured into the host office control program (HOCP) and the distributed office control program (DOCP).

Electronic Mail Configuration

An electronic mail configuration is a differentiated system that combines document access and distribution and computer-based message capability by combining hardware and software elements in the host, the distributed computer, and local workstations. An electronic mail system has a configuration identical to the information storage and retrieval configuration (Figure 8-10) and is commonly built on a foundation of a word processing configuration and computer networks in a distributed environment. As covered in Chapter 7, storage facilities for documents and messages can reside in any of the control centers and can be accessed from associated workstations.

Other Configurations

With the integration of voice and image data into office information systems, it is expected that systems will become more and more highly distributed. Local intelligence is needed to process voice and image data, and more storage capacity is needed in the distributed computer system. Advanced features such as pattern and voice recognition require extensive processing power, so that the influence of the host computer on the total configuration is expected to remain strong in future years.

Figure 8-10. Information storage and retrieval system.

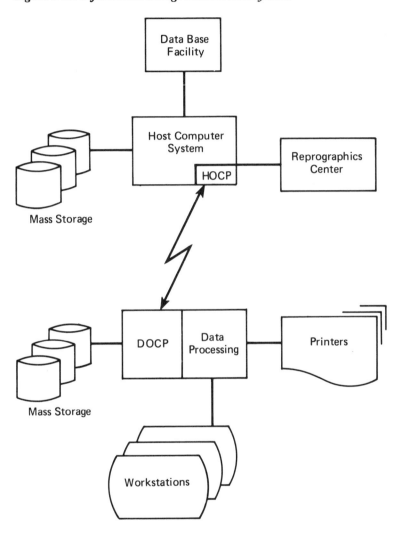

Key: HOCP = Host Office Control Program
     DOCP = Distributed Office Control Program

Packet Switching

A technical area that is of primary concern in distributed office systems and computer-based message systems is the characteristics of the network used for telecommunications. This section gives a technical overview of packet switching and can be skipped without a loss of continuity in the subject matter. Packet switching is a key communications issue in some European countries, and it is an area in which notable successes have been achieved in the United States in recent years.

Overview

The selection of an appropriate class of service for a particular network is a complicated process that involves a complete analysis of call duration, distance, data rates, and other business issues. Connect times and traffic volumes must be matched with classes of service to achieve an operational window for specifying service requirements. Backup facilities are also problematic, since potentially unreliable multipoint connections are commonly used to reduce line costs. An overall compromise solution to the establishment of near-optional networking facilities—at least conceptually—is packet switching, which combines the flexibility of circuit switching with the economics and reliability of leased lines with point-to-point connections.

Classes of Service

There are several classes of service relevant for a discussion of packet switching:

Leased lines.
Dial-up service.
Message switching using circuit-switched lines.
Packet switching using public network facilities.

Each class of service has its unique advantages. *Leased lines* are suitable for bulk-data operations, such as file transfer, but can be inefficient for low communications loads. *Dial-up service* is suitable for low-volume usage, but connection time can be excessively long.

Message switching combines the convenience of leased lines with the economy of dial-up service but suffers from excessively long response time. *Packet switching* provides good response times with economical costs and is suitable for facilities that have relatively low utilization rates but require fast response times. The CCITT definition of packet switching is "the transmission of data by means of addressed packets whereby a transmission channel is occupied only for the duration of the transmission of the packet."

Comparison of Message Switching and Packet Switching

In both message switching and packet switching, the routes from origin to destination for communications traffic can be conceptualized as passing through several intermediate computer nodes in the network. Messages are stored at one of these nodes and are then transmitted on to their destination. While this mode of operation exists for both message and packet switching, there are significant differences between the two modalities.

Message switching is intended for non-real-time traffic between people, with delivery times in the neighborhood of "fractions of hours." Messages are often long and stored in their entirety at each intermediate node. Moreover, messages can be stored permanently at network nodes for future retrieval.

With packet switching, the response time is designed to be a fraction of a second, and messages are transmitted as fixed-size blocks called packets. Packets are not stored permanently, but only as long as needed to ensure accurate and reliable transmission.

Message and packet switching also differ structurally. In message switching, there is normally a dominant center (or *node*, as it is called), whereas in packet switching, the network has an amorphous structure.

Figure 8-11 depicts a comparison between circuit switching and packet switching, and Figure 8-12 conceptualizes the differences between message switching and packet switching.

Routing

In packet switching, a general method of routing is normally used to increase reliability. On the way from origin to destination,

Figure 8-11. Comparison between circuit switching and packet switching.
(*Source: "Principes de la commutation par paquet."*)

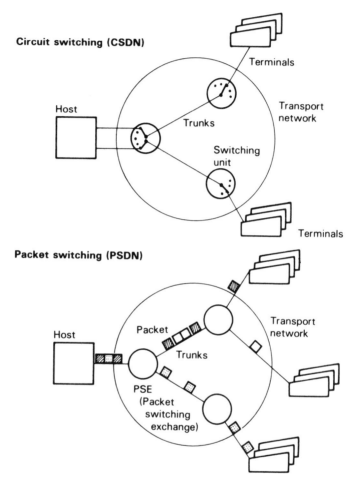

a packet is stored temporarily at each intermediate node. When the time for transmission to the next node occurs, the corresponding network computer selects a "best choice" route based on availability, traffic volume, congestion, and optimum path length. Adaptive routing of this type is characteristic of packet switching systems.

Figure 8-12. Conceptual differences between message switching and packet switching. (Source: "Principes de la commutation par paquet.")

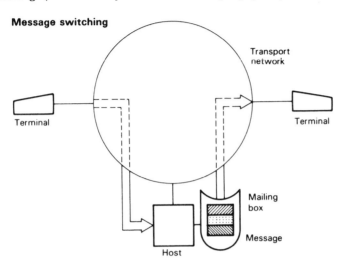

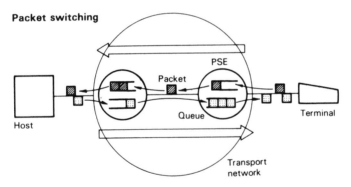

Types of Packet Switching

Two types of packet switching exist: virtual circuit service and datagram service. With *virtual circuit service,* a logical connection must be made between the origin and the destination of a message. With *datagram service,* a packet is entered into the network together with a specification of its destination, and the packet switching network establishes a connection dynamically.

Virtual circuit service may involve establishing a logical connection between origin and destination for each message, which in fact may be composed of several packets, or it may take the form of a preestablished logical circuit eliminating the need for a logical connection each time a message is delivered into the network. The former type of service is called a *virtual call* and the latter is known as a *permanent virtual circuit.*

Protocols

A protocol is a convention for interfacing data terminal equipment (DTE) with data communications equipment (DCE). Front-end processors and terminal controllers are classed as data terminal equipment. Network computers are data communications equipment.

Standard protocols for various aspects of data communications have been established by the International Standards Organization (ISO). The best-known protocol of this type is known as X.25, which is used to manage a communications session.

The ISO is currently developing a model of a data transport network, which is shown in Figure 8-13. Each of the model's seven layers is concerned with a particular aspect of communications, and each layer makes use of the one below it. The X.25 protocol is concerned with the basic requirements for interconnection. It covers the three lowest layers of the ISO model, which are the only ones that will be explained here. These three layers are:

- Layer 1—physical layer. This layer defines the physical, electrical, functional, and procedural characteristics needed to establish, maintain, and disconnect the physical link between the DTE and the DCE.
- Layer 2—frame or data link layer. This layer describes the processing performed by software in the DCE and DTE to maintain control over data transmission, error checking, and packet handling.
- Layer 3—packet or network layer. This layer defines the packet format and control procedures for exchanging packets between the DTE and the DCE.

Figure 8-13. ISO model of a data transport network.

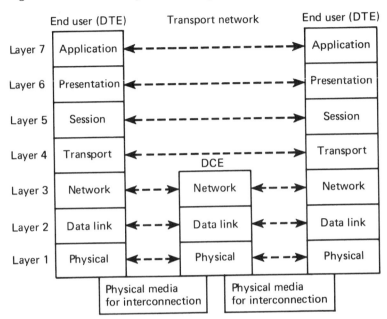

When packet transmission is performed, layer 3 facilities utilize layer 2 facilities, which in turn utilize the physical aspects of layer 1. Each horizontal arrow in Figure 8-13 represents a protocol, but the entire interfacing process is also referred to as a protocol.

Higher layers in the ISO model of a transport network are concerned with access to the network by an end user and with applications.

Intersystem Connection

A facility designed to connect two networks is known as a *gateway*. A gateway is concerned with interface standards and addressing schemas. Specifications have been established for network interconnections (X.75) and for intersystem addressing (X.121); they are implemented through a network computer that manages traffic between networks through flow control and buffering procedures.

User Benefits

There are many benefits to be obtained by using the packet switching method. These benefits are generally summarized by the following list:

Lower costs—the user pays only for the periods of activity.
High reliability—backup elements are built into the packet switching system.
High accuracy—error checking occurs at the access and switching levels.
High accessibility—permits intersystem connection between diverse public networks.

Certainly, packet switching technology is not applicable to all networking problems. When it is applicable, however, the benefits listed, as well as the administrative advantages given in the following section, can be realized.

Administrative Benefits

Some of the administrative benefits available through the use of packet switching technology directly or indirectly relate to user benefits such as cost and reliability. These benefits invariably possess a technical basis but reflect upon the "management" aspect of networking:

International standards—standardized protocols and procedures facilitate network development and maintenance.
Worldwide coverage—the capability for connecting remotely is available, a capability that would not be possible through private cooperation.
Resource sharing—a high level of functional capability is available that would not be possible with private facilities.
Network management—the packet switching facility establishes network management capability that provides for convenience and reliability.
Security—closed subscriber groups permit a high level of basic network security.
Performance—by averaging the needs of diverse users, the packet switching service can provide a relatively high level of

throughput that is independent of the distance between users.

Growth—eliminating the need for a user to specifically plan for major increases in traffic volume.

The basis for worldwide packet switching service is the X.25 set of protocols developed in conjunction with the CCITT and other agencies for standardization. As such, X.25 is a recommendation for international cooperation and exists as a point of reference in telecommunications technology.

Local Communications

Local communications facilities are intended to connect data terminal equipment—computers, workstations, reprographics centers, and so forth—that is located in a physically restricted area, such as a building or an office complex. A system of data terminal equipment and local communications facilities is called a *local network*. Local networking capability serves as a low-cost alternative to the omnipresent telephone system.

Problem Domain

The objective of a local network is to connect every station with every other station, as in the case of connecting a stand-alone word processing system with an intelligent copier or an image transfer device to a remote computer. The management of a local network is decentralized, which means that no station is in control.

In a local network configuration, the distances between stations are important for both the functionality and the cost of the transmission facility. Local networks are also easily expanded, without upsetting the system, by adding stations.

Local Network Topology

There are four types of local networks: bus structure, ring structure, star structure, and an unconstrained graph network. Bus and ring structures are the most prevalent and are discussed here.

Figure 8-14 depicts both bus and ring local networks. A *bus* is a cable to which stations are attached. When one station transmits, signals branch out to both ends of the bus, reaching other stations. A *ring* is a circular cable connecting a set of repeaters. Stations are attached to repeaters. Once a station transmits, the repeater places the information in the ring, where it is passed to the destination repeater and station. Ring networks can span greater distances because repeaters are used.

Contention Methods

Contention methods of traffic management in local networks are most common in bus structures and in light to moderate traffic load. The most widely known contention method was developed and patented by the Xerox Corporation and used in its Ethernet bus system. The method, named Carrier Sense Multiple Access with Collision Detection (CSMA/CD), uses two rules to control the operation of the network:

Rule 1:   When the channel is being used by a station, all other stations must wait before attempting to transmit their own messages.

Rule 2:   When the channel is clear and two stations attempt to transmit at the same time, a "collision" occurs. When this happens, both stations quiesce and attempt a transmission at a later time.

The contention method is an efficient technique: studies have shown that stations have a transmission success rate greater than 99 percent on their first attempt.

Slotted Rings

A slotted-ring technique is used with ring structures to control access and manage traffic. A set of fixed-length frames (called packets or slots) is circulated through the network. When a station wishes to transmit, it waits for the first empty slot and places its message in it along with a destination. When the message arrives at its destination, the receiving station copies the message and marks the slot "acknowledged." When the slot finally gets back to the

Figure 8-14. Common local network structures.

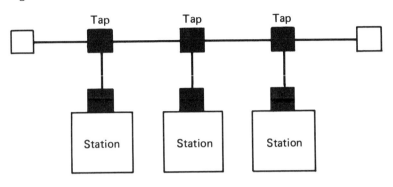

Bus Network

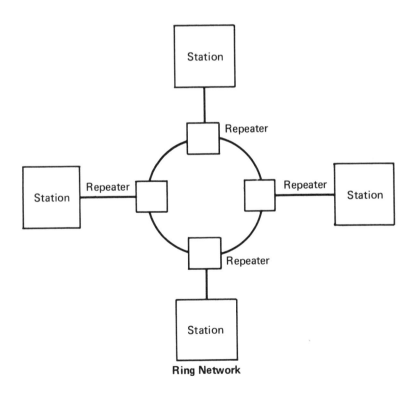

Ring Network

sender, the sender senses the "acknowledged" indicator and marks the slot empty. A station may not immediately reuse a slot. Two advantages of the slotted-ring technique are that it is efficient under heavy loads and that it is self-acknowledging.

Token Passing

A widely used method of controlling contention in a ring network is the use of a token—either a byte or a word of fixed information—that circulates during idle periods. A station wishing to transmit waits until the token comes by and interrupts it. The station then transmits its message and subsequently recreates the token and circulates it. Thus a station cannot transmit unless it has taken the token, and the token circulates only when the network is free. This method allows variable-length messages and high network utilization.

Bandwidth

The amount of data that can be transmitted per second is known as bandwidth. Base-band systems permit text and data transfer. Broadband systems additionally permit voice and image data to be combined with text and data. Cable television, microwave, and fiber optics are other techniques that are applied to local communications and offer broadband capability.

Digital PBXs

An in-house telephone switching system that permits direct inward and outward dialing and extension service is known as a *Private Branch Exchange* (PBX). Traditionally, analog PBXs have permitted data traffic between extensions for local communications, but modems were needed on either side of the PBX. This has been a serious objection to their use on an in-house basis for computer traffic.

Digital PBXs can be used with computer systems without modems and provide an effective local communications medium. For voice traffic, analog waveforms are sampled 8,000 times a second before conversion to a binary pattern.

Summary

The systems approach to office systems emphasizes the objectives and functions of a system, recognizing the interrelationship between pages, information, and computers. In designing office systems, as much precision as possible is desirable. It is useful, therefore, to view a system along four dimensions: congruence, integrity, auditability, and controllability. On the basis of internal structure, systems can be categorized as segmented or differentiated.

In office systems design, an interoffice system is known as an organizational network, whereas an intraoffice system is regarded as an activity network. Within the domain of organizational and activity networks, a "switching center" approach is used for functionality and for cost considerations. Five basic systems configurations were identified:

Word processing.
Text/data integration.
Advanced text processing.
Information storage and retrieval.
Electronic mail.

Packet switching is a computer networking technique that lends itself to office systems and exists as an alternative to dial-up and leased telephone service. Packet-switching protocols are based on international standards and have potential for widespread acceptance.

Local communications facilities are intended to link data terminal equipment that is located in a physically restricted area, such as a building or office complex. Two basic kinds of local network topology are commonly used: bus structure and ring structure. Bus networks are predominantly controlled by contention methods. Ring networks use slotted rings and token passing for control. Digital PBXs also exist as a convenient method of implementing local communications.

SUGGESTED READING

Davies, D. W., D. L. A. Barber, W. L. Price, and C. M. Solomonides, *Computer Networks and Their Protocols*, New York: Wiley, 1979.

Katzan, H., *Systems Design and Documentation: An Introduction to the HIPO Method*, New York: Van Nostrand Reinhold, 1976.

Sutherland, J. W., *Systems: Analysis, Administration, and Architecture*, New York: Van Nostrand Reinhold, 1975.

CHAPTER 9

Management Strategy

A recent ad in a major newspaper reads: "Computerize now or fall behind! Modernize with a Brand X computer. It's easy and affordable!" Unfortunately, things aren't that simple. There are many potential opportunities to be derived from office automation, but almost as many pitfalls to be avoided. Since it is impossible to find a path that fully exploits each opportunity and avoids every pitfall, a balanced approach is needed. This chapter examines management's role in establishing such a balanced approach that sets a clear direction and weighs benefits and resources.

Concept of Strategy

The measure of a manager's success is whether he or she moves the business to a better position than the starting point. The process of determining where the business stands is conveniently called a *situation audit.* The objective of a situation audit is to identify and analyze key trends, forces, and other phenomena having a potential impact on the formulation and implementation of strategies. This step is critical to the development of a successful strategy.

It is of course impossible to analyze every item of information in the operating environment. Management must identify in the changing environment those items that have the greatest importance to the strategy being developed. The situation audit systematically assesses environmental impacts and pulls together the di-

143

vergent views of different parts of the organization. The situation audit is the starting place for the specification of objectives and the delineation of strategy.

Definition of Strategy

Strategy is concerned with the manner in which management chooses to use an enterprise's resources to reach its objectives. Thus a strategy is a relationship among the enterprise, its objectives, its resources, and its environment.

In business, strategies are normally grouped into two categories: business strategies and functional strategies. Business strategies are concerned with the coordinating activities that relate the various functions of a business to the business objective. Functional strategies, on the other hand, coordinate activity within a business function. Automation activity, for example, is traditionally covered by a functional strategy, since in most cases it is not viewed as a profit center or other unified business function.

Effective strategies aid management in directing attention at the important issues. Hierarchies of strategies naturally correspond to hierarchies of management. A lower-level strategy takes as its objective one part of a broad course of action that the next higher level of strategy calls for.

Statement of Strategy

A statement of strategy is primarily a vehicle for focusing attention on the strategic aspects of the enterprise. It is a means of communication for those who must review and approve the strategy, gain guidance for their activities, maintain the strategy for making adjustments to it, or use it to measure performance. The elements of a strategy then use the objectives and environmental factors that have been identified. They incorporate the following elements:

> *A broad course of action*—what is the general direction to be followed in attempting to achieve the stated objectives?
>
> *Assumptions*—what are the major assumptions on which the strategy is based?

Risks—what is the nature of the risks in the strategy, and what is their potential impact?

Options—what options are available within the strategy, how long are they good for, and what are the considerations on which a selection should be based?

Dependencies—what are the key dependencies of this strategy, what is their nature, and in what way are they significant to the strategy?

Resources required—what are the resources required to carry out this strategy? Does it require any unique type of resource?

Financial projection—what is the financial projection of costs and benefits?

Alternatives—what alternatives were rejected in the selection of the strategy? Briefly, what were the reasons for rejection?

Because office automation is evolutionary in most organizations, strategy precedes planning. The concepts are related. In fact, many people would use the term "strategic planning" as a good substitute for "strategy."

Development of a Strategy

The development process for strategy is one area where strategy differs from planning. Not all strategies are analogous to planning; some strategies reflect a course of action for a particular problem situation, while others may be stand-alone strategies to support or develop one particular aspect of a plan.

The steps to be followed in developing a strategy are:

1. *Understand the scope and nature of the area of concern.* Take ample time in gaining an understanding of the area of concern to arrive at a concise, sharply focused statement of it.

2. *Describe the future environment.* Develop in reasonable depth a broad understanding of the future environment that influences the area to be studied.

3. *Identify the objectives and the alternative possible strategies.* Objectives for strategic plans are set and, if necessary, modified according to the regular planning process. Functional strategies may need to identify several possible objectives, and then select one after testing them for credibility and attainability.

4. *Set up criteria for selection.* In order to exercise a reasonable choice among the alternative strategies, select the criteria for using them. The criteria should measure the basic long-range effects on the company of following an alternative.

5. *Select the preferred strategy and prepare the statement of strategy.* Select the best strategy by using the criteria to measure the effect that could be expected from selecting each alternative. Then develop the selected strategy in greater detail and prepare a statement of it.

6. *Get the strategy approved.* For functional strategies, get appropriate line and staff review and approval of the statement of strategy and its supporting data, reasoning, and other forms of evidence.

7. *Incorporate a functional strategy into the strategic and operating plans of the organizations that are involved.* The process of incorporation will require changes in the "Objectives" and "Strategy" sections of these plans, and in the appropriate business and functional area plans.

So much for the topic of strategy. It is important to emphasize, however, that a strategy for office automation is not an end in itself. The practice of incorporating an office automation plan into an overall business plan is precisely what is needed to ensure that you are dealing with reality and not a world of wishful thinking.

Office Automation Strategy

Because office automation is an evolving technology, the key question is not what should be done tomorrow but rather what should be done today to prepare for an uncertain future. Thus the dimensions of an office automation strategy are threefold: direction and goals, a framework for action, and a rationale for decision making. A strategy does not specify individual applications, justification, or a definition of the inherent technology.

Contents of the Strategy

An office automation strategy gives three things: where we are (*current position*), where we are going (*goals*), and how we get there (*direction*). *Current position* is a specification of the equipment, already installed applications, the trained and knowledgeable people, and existing organizational problems that have a bearing on an office automation plan. *Goals* are certainly dependent on the particular enterprise but include factors such as better customer service, increased sales volume with the same head count, reducing administrative expense, job enhancement, and the establishment of new markets.

The *direction* (or "How do we get there?") is a major issue—in fact, it is the reason for a strategy in the first place. Direction needs policies and procedures in the following areas: justification, implementation, user acceptance, and staffing and organization. These areas form the basis for a strategic plan that gives the stages of tactical (or functional) planning for the enterprise.

Strategic Plan

The strategic plan for office automation covers three stages of work: preparatory work, development of the tactical plan, and use of the tactical plan.

The preparatory work is crucial because it sets the stage for success or failure. The history of planning in the enterprise should first be reviewed, to ensure that the office automation strategy will fit smoothly into the larger business plan. High-level management commitment—or a sponsor—is needed to kick off an effective project. Responsibility for office automation and also for tactical planning should be assigned, and a strategy group should be formed. Participation in any of these activities need not be on a full-time basis; the key is to specify basic objectives and output of the strategy sessions.

The sponsor is particularly significant because resources have to be obtained for office automation, even for planning itself, and the needed policies must be set and enforced.

In developing the plan, assumptions must clearly be made concerning the organizational structure, the staff, and the relevant technology. After gathering and analyzing information about the organization, the strategy group selects a number of office automation applications to be implemented. Objectives are required at this stage, including opportunities to be exploited and measurable goals to be achieved. The strategy group must also establish implementation schedules and identify any problems that are anticipated. The strategy plan, at this point, should be reviewed with local management in the areas of data processing, communications, administration, facilities, and personnel and then presented to top management.

In using the plan, approval is requested for initial activities, including pilot projects, research efforts, and associated development work. Means of implementing the plan are also needed, including people, procedures, and a user feedback channel. The plan should also include a list of action items to be updated in line with the planning policies of the enterprise.

Guidelines

In outlining a strategy and developing a plan for office automation, a few pertinent guidelines are helpful for increasing the chance of project success. The most important consideration is to concentrate resources in time and in place and focus these resources on the stated objectives. Second, it is not prudent to allocate resources unless there is a better-than-average chance of success. There are simply too many intangibles in office automation to ensure immediate success, and user acceptance can be problematic unless it is handled properly. Naturally, due consideration of user acceptance is an important aspect of strategic planning. Finally, it goes almost without saying that you shouldn't compete for the organization's resources unless your organizational unit is likely to receive them and unless they will help your unit achieve its strategic objectives.

Justification

Enthusiasm over office automation is contagious. The technology of word processing and electronic mail, for example, is exciting. It is likely that this enthusiasm will carry over to the executive suite, where seasoned executives will recognize the glamour of the newly discovered source of productivity. In any case, justification will be required regardless of the credibility and influence of the sponsor.

Definition

Justification is the information presented to a decision maker to support an investment proposal. In general, there are two major reasons to provide justification: to achieve agreement to put an investment proposal in an overall plan for the organization, and to obtain a commitment of resources for an implementation project. In the area of office automation, justification is notably difficult, since hard money is being balanced against soft savings.

Method

Traditional methods of justification include almost any form of inductive reasoning. Some of the more noteworthy are:

Cost of automation is lower than that of the manual method.
Automation improves customer service.
There is no other way to do it.
Automation gives a competitive advantage.
It is part of the cost of doing business.
It is our "image" to be at the leading edge.

With office automation, a useful approach is to show the labor cost curve with its flattened slope achieved by increasing the efficiency of the affected personnel.

Productivity

Productivity is a key benefit of office automation, becuse it provides an increased quality and quantity of work, an increased span of control, more timely work, and decreased turnover of key people. Thus productivity can be more formally specified as the relationship between the output of a work unit and its input in labor and raw materials.

Methods of Analysis

Two methods of delineating productivity are the payback method and the cash flow method.

The payback method reflects costs that are displaced because of office automation technology. To use videoconferencing as an example, travel costs are reduced through meetings and conferences conducted with the latest methods in video technology. Clearly, a certain percentage of travel costs is replaced per year, and after a period of years, the initial investment in equipment pays for itself.

With the cash flow method of analysis, costs are not replaced by the newer technology, but they are reduced. Analog versus digital private branch exchange serves as an example. As the level of data communications increases through an analog PBX, costs increase because more modems are needed. With a digital PBX, the initial investment in the equipment is higher than for an analog PBX, but the additional costs associated with increased data communications are lower.

In some cases, therefore, labor, equipment, and facilities costs are displaced and *cost savings* result. Typical examples are floor space for files, travel costs, postage costs, forms and supplies costs, and time savings. In other cases, *cost avoidance* takes place because there is decreased growth of staff, equipment, and facilities.

Implementation

Implementation refers to a pilot project to establish a concept and then an expansion both horizontally and vertically within the organization. In this context, horizontal expansion refers to a proliferation of the same application throughout the organization. Vertical expansion refers to an enhancement of the application domain to add new functions.

The Pilot-Project Approach

A pilot project is a forerunner of a larger project or application with the objective of building confidence in a new idea. A pilot project is a learning experience that demonstrates technical feasibility, costs, benefits, and user acceptance. Effectively, a "pilot" is a means of planning big and starting small.

Clearly, a pilot project is an approach to experiential learning and to gauging user resistance. Some approaches that have been taken are to give the office automation function to some group and not others, or to give the function to all groups, and then take it away. The objective in both cases is to measure productivity, take interviews, and give questionnaires to assess the success of the project.

In many cases, and particularly in office automation, people simply do not know what they want and what they need. A pilot project is a means of getting valuable input to the planning process without spending an excessive amount of money. It is also a means of finding mistakes and problems early and thereby minimizing exposure. If a pilot project flops, it is chalked up to experience. If a heavily committed project fails, the organization may be seriously hurt.

Selecting a Climate for Success

The best pilot project is naturally one with a high probability of success. The object system should be easy to use and easy to learn. It should be assigned to a relatively small, close-knit group with good internal communication, so its members can aid one another, and with an enthusiastic manager. Most important, a pilot project must not be placed in a pressure group that may not give a concept a fair evaluation.

The duration for a pilot project must be long enough for the people to become accustomed to the system but short enough so the momentum and enthusiastic atmosphere do not subside. Most experts agree that two to six months is the optimum duration.

A group of users who are currently using terminals is a good candidate for a pilot system. It is also important that the people in the selected group recognize that they need improvement. While these factors are not always possible, they greatly increase the chances of success.

Choosing the Right Application

As long as office automation is dependent upon people, it is necessary to involve people in the pilot project. If a person can achieve personal success with a system, then he will remain the best source of support.

Some office automation applications are people-intensive; word processing, communicating word processors, and computer-based message systems are good examples. Other applications, such as local networks, digital PBXs, and facsimile systems, are technology-driven and do not provide the high degree of user feedback needed for project evaluation. Clearly, these other areas may be extremely valuable to the enterprise, but must be cost-justified using traditional means.

User Acceptance

Office automation is a change agent, and the success of a project is largely dependent upon the user's reaction to it. It is important that the people react to the implementation phase with: "This is our system." Much of change management is dependent upon an acceptance strategy, resistance management, and proper education and training. It is important for managers to remember that no matter how carefully an office automation system is chosen, it can be undermined by employees who resent it.

Acceptance Strategy

A successful acceptance strategy is anticipatory: fears, resistance, and expectations must be identified early in the implementation plan. The announcement of the plan should be made early on in the change cycle, and significant policy questions have to be addressed.

The system announcement should in general be made to the affected department by a high-level executive, giving the objectives, benefits to the individual and the organization, expectations, feedback mechanism, and implementation schedule.

The acceptance strategy is a link between the strategic and tactical plans of an enterprise and covers personnel policies regarding relocation and terminations, associated job levels, training plan, and requirements for using the office automation equipment. A fairly widespread relocation policy is that people not wanting to do office automation will be transferred within the organization, but new employees coming into the affected department must use it.

Resistance Management

Some of the factors that contribute to a reduced resistance are also good management practices in general. Probably the most significant aspect to consider in this area is the human factors of the hardware, software, and work environment. Ample free "terminal" time is an absolute must, and no worker statistics should be collected. A "help" desk, already available in many data processing departments, is useful for easing tensions but also as a window into the functionality of the system.

Education and Training

The existence of training programs in modern organizations is now taken for granted. Training is supplied by practically all vendors and in-house training groups.

Best results are achieved when the subject matter is presented in the following sequence: overview, a basic-functions course, an advanced-functions course, and tutorials interspersed whenever necessary. On-the-job training is recommended between basic and advanced training.

Poor performers should be identified early in order to provide for additional assistance or personnel action. Final evaluations for performance standards and certificates of completion are mandatory.

Training should be timed with the acquisition of the system. If training is performed too early, the employees lose their enthusiasm and confidence, and the chances of user acceptance are lessened.

Figure 9-1. Organization structure based on a data processing orientation.

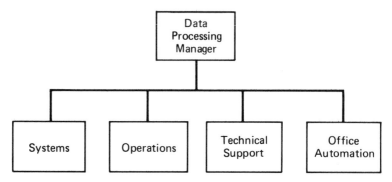

Staffing and Organization

Two major aspects of office automation remain to be covered from the viewpoint of management strategy: staffing and organization. In the people category, both job function and personal qualifications must be regarded. Organization relates to where office automation is placed in the total organization and the missions of the various organizational units. Organization precedes staffing in this section because organizational structure precedes missions and so forth.

Organizational Structure

The reporting structures in an enterprise are based on the history of the enterprise, its people, and its present organizational structures. The most common form of organizations is given in Figure 9-1, which reflects a *data processing orientation*. The advantage of this form is that the management and the staff understand the technology. A disadvantage is that the staff may be spread too thin for an additional mission.

Figure 9-2 gives a form of organization with an *information systems orientation*. The information systems manager has the general mission of managing information, with no other responsibilities. This exclusivity is a major advantage. Figure 9-3 gives an or-

Figure 9-2. Organization structure based on an information systems orientation.

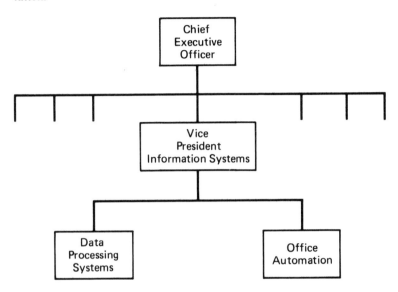

ganization structure based on a *communications orientation*. It is analogous to the data processing form, with the same general advantages and disadvantages.

Figure 9-4 depicts an organization structure based on a *management services orientation*. This form may have lesser problems with user acceptance but may require heavy support from other units.

Basic Missions

The mission of the office automation department is to make information services available to the user departments. Regardless of the form of organization, the following tasks are relevant:

Planning.
Implementation.
Support.
Education.
Management of information.
Control and operations.

Figure 9-3. Organization structure based on a communications orientation.

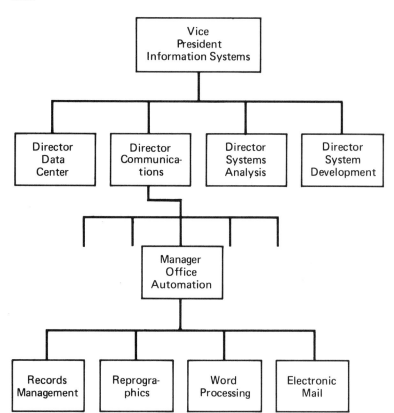

Job Functions

The job functions pertinent to a wide cross section of office automation groups are:

Strategic planning.
Tactical planning.
Office automation planning.
Office automation specialists.
Text administration.

Figure 9-4. Organization structure based on a management services orientation.

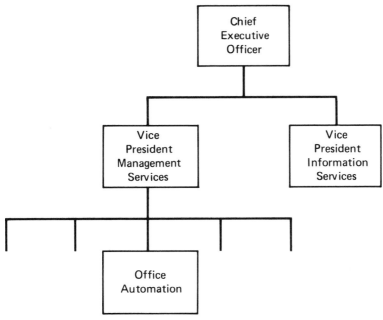

Office automation education.
Coordination and information control.

Strategic and tactical planning categories are usually considered higher-management functions and probably will not be full-time positions. A *steering committee* composed of members from related functional areas (such as data processing, communications, personnel, and so forth) guides the action and directs the development of the strategic plan. In some organizations, tactical planning can be done by the staff of the office automation department. Otherwise, each of the job functions is normally organized along the lines of office automation applications. For example, there is management associated with word processing, electronic mail, and so forth. Similarly, there are also specialists and coordinators in each of these areas.

Staff Requirements

Staff requirements are not unique in office automation, except for the office automation specialists, who need a wide range of experience and skill. Some relevant qualities are:

- Must be able to work with user groups in identifying requirements for office automation.
- Must be able to conduct training sessions and build user support.
- Must be able to conduct demonstrations.
- Must be able to objectively evaluate equipment and disseminate information to relevant organizational units.
- Must understand the operation of the organization.
- Must have political savvy.
- Must understand computers and data processing.
- Must be able to pursue self-education.
- Must be able to communicate clearly both orally and in written form.
- Must be able to interface with personnel with differing backgrounds.

In the near future, the source of most office automation specialists will be the data processing department. In the distant future, the integration of data processing, office automation, and communications will lead to a need for a new breed of "information specialists."

Summary

Because of the many different opportunities that confront an executive, a balanced approach to office automation is needed, starting with a situation audit and continuing through a strategic plan.

An effective strategy for using an enterprise's resources is needed; it is customarily broken down into two areas: business strategies and functional strategies. In order for a strategy to be a vehicle for planning, a statement of strategy is needed. Fortunately, the notion of strategy is well defined and the steps in the development of a strategy can be delineated.

An office automation strategy starts with a current position and yields goals and directions. The direction is used to establish policies in the following areas: justification, implementation, user acceptance, and staffing and organization.

In strategic planning a sponsor is particularly useful, because hard resources are needed for soft results. Guidelines for implementing an office automation plan are extremely useful for increasing the chances of project success.

Justification involves presenting information to a decision maker to support an investment proposal. In this effort, methods of justification, productivity assessment, and methods of cost–benefit analysis are important.

Implementation starts with a pilot project to support the office automation concept. A pilot project is a forerunner of a larger project or application, with the objective of building confidence in a new idea. Significant considerations are creating the right climate for success and choosing the appropriate application.

User acceptance is the primary ingredient in a successful project. Gaining user acceptance requires a strategy of its own. Resistance management, education, and training are considerations.

Staffing and organization refers to organization structure, basic missions, job functions, and staff requirements. Lists of desirable characteristics in each category should be compiled to provide guidelines for management action.

SUGGESTED READING

Bingham, J. E., and G. W. P. Davies, *Planning for Data Communications,* London: Macmillan, 1977.

Buehler, V. M., and Y. K. Shetty, *Productivity Improvement: Case Studies of Proven Practice,* New York: AMACOM, 1981.

Layard, R., *Cost–Benefit Analysis,* Baltimore, MD: Penguin, 1972 (Chapter 6: "The Value of Time").

McLean, E. R., and J. V. Soden, *Strategic Planning for MIS,* New York: Wiley, 1977.

Nolan, R. E., R. T. Young, and B. C. DiSylvester, *Improving Productivity through Advanced Office Controls,* New York: AMACOM, 1980.

Steiner, G. A., *Strategic Planning: What Every Manager Must Know,* New York: The Free Press, 1979.

Bibliography

Abraham, S. M., "Is Office Automation the Best Darned Thing You've Ever Seen? MAYBE," *Computerworld*, September 28, 1981, p. 21.

Bailey, A. D., J. Gerlach, R. P. McAfee, and A. B. Whinston, "Internal Accounting Controls in the Office of the Future," *Computer*, May 1981, pp. 59–70.

Ball, L., "Data in WP: Very Valuable—And Very Easy to Steal," *Computerworld*, September 28, 1981, p. 53.

Berger, V. L., "WP Service Bureaus a Way to Handle Peak Loads Without Having to Buy Additional Equipment," *Computerworld*, September 28, 1981, p. 60.

Bingham, J. E., and G. W. P. Davies, *Planning for Data Communications.* London: Macmillan, 1977.

Blumenthal, M., "Office Automation Market: Too New to Call," *Computerworld*, September 28, 1981, p. 35.

Boyer, R. D., *Computer Word Processing: Do You Want It?* Indianapolis, IN: Que Corporation, 1981.

Buehler, V. M., and Y. K. Shetty, *Productivity Improvement: Case Studies of Proven Practice.* New York: AMACOM, 1981.

Canning, R. G., "The Automated Office: Part I," *EDP Analyzer*, September 1978.

Chorafas, D. N. *Office Automation: The Productivity Challenge.* Englewood Cliffs, NJ: Prentice-Hall, 1982.

Connor, U., "Success of Office Automation Depends on User Acceptance, Not High Technology," *Computerworld*, September 28, 1981, p. 46.

Cypser, R. J., *Communications Architecture for Distributed Systems.* Reading, MA: Addison-Wesley, 1978.

Dainoff, M., "What Price Comfort? The CRT Terminal in the Office," *Computerworld*, September 28, 1981, p. 15.

Data Security Controls and Procedures: A Philosophy for DP Installations. White Plains, NY: IBM Corp., Form G320-5649, 1977.

161

Dertouzos, M. L., and J. Moses, *The Computer Age: A Twenty-Year View.* Cambridge, MA: MIT Press, 1979.

Doll, D. R., *Data Communications: Facilities, Networks, and Systems Design.* New York: Wiley, 1978.

Driscoll, J., "Is Office Automation the Best Darned Thing You've Ever Seen? NO," *Computerworld,* September 28, 1981, p. 20.

Drucker, P. F., *Managing in Turbulent Times.* London: Pan Books, 1980.

Edwards, M., "Electronic Mail: Something for Everyone," *Infosystems,* March 1981, p. 54.

Edwards, M., *Office Automation Primer for Federal Office Administrators.* Wayland, MA: Federal Office Institute, 1981.

Evans, C., *The Micro Millennium.* New York: Viking, 1979.

Finn, N. B., "Office of Future Seen Answer to Paper Blizzard," *Computerworld,* September 28, 1981, p. 71.

Fohl, M. E., *A Microprocessor Course.* Princeton, NJ: Petrocelli Books, 1979.

Folts, J. R., "WP/DP: Getting Married . . . Or Just Living Together?" *Computer Decisions,* July 1981, p. 44.

Gaffney, C. T., "Crisis in the Work Place: Selling the Staff on Office Automation," *Computerworld,* September 28, 1981, p. 11.

Galitz, W. O., *Human Factors in Office Automation.* Atlanta, GA: Life Office Management Association, 1980.

Goldman, M., "Guerrilla Tactics," *Computerworld,* September 28, 1981, p. 7.

Hansen, J. R., "How Small Computers Carry the Mail," *Small Systems World,* November 1980, pp. 21–23.

"Hilton to Offer Videoconferencing Throughout the Hotel System," *Hilton Items,* Beverly Hills, CA: Hilton Hotels Corporation, November 1981, pp. 1–2.

Horrigan, J. T., "Workers Said Fearful of Creeping Automation," *Computerworld,* September 28, 1981, p. 68.

Housley, T., *Data Communications and Teleprocessing Systems.* Englewood Cliffs, NJ: Prentice-Hall, 1979.

IBM Displaywriter System: General Information Manual, Austin, TX: IBM Corp., Form G544-0851, 1981.

IBM Distributed Office Support Facility: General Information. Kingston, NY: IBM Corp., Form GC27-0546, 1981.

Inose, H., "Communications Networks," *Scientific American,* September 1972, pp. 117–128.

Jones, R., "Food Firm Finds Office Automation Fortifying," *Computerworld,* September 28, 1981, p. 69.

Katzan, H., *An Introduction to Distributed Data Processing.* Princeton, NJ: Petrocelli Books, 1979.

Katzan, H., *Distributed Information Systems*. Princeton, NJ: Petrocelli Books, 1979.

Katzan, H., *Introduction to Computers and Data Processing* (Chapter 25: "Automated Offices"). New York: D. Van Nostrand, 1979.

Katzan, H., *Invitation to FORTH*. Princeton, NJ: Petrocelli Books, 1981.

Katzan, H., *Operating Systems: A Pragmatic Approach*. New York: Van Nostrand Reinhold, 1973.

Kerr, S., "Westinghouse Electric's Pilot Program Grows into Global Network," *Computerworld*, September 28, 1981, p. 50.

Kinnucan, P., "Local Networks Battle for Billion-Dollar Market," *High Technology*, November/December 1981, pp. 64–72.

Kraft, P., "Mid-Level Managers: Will They Fade Out As Automation Changes the Office?" *Computerworld*, September 28, 1981, p. 29.

Layard, R., *Cost–Benefit Analysis*. Baltimore, MD: Penguin Books, 1972.

Manz, B., "WP Now Brings Message Center to Heart of Office Environment," *Computerworld*, September 28, 1981, p. 74.

Martin, J., "Successful Office Automation," *Computer Decisions*, Part I, June 1981, p. 56; Part II, July 1981, p. 108.

Maryanski, F., "Office Information Systems," *Computer*, May 1981, pp. 11–12.

McLean, E. R., and J. V. Soden, *Strategic Planning for MIS*. New York: Wiley, 1977.

McQuillan, J. M., *Electronic Mail: The Message System Approach*. Cambridge, MA: BBN Information Management Corp., 1981.

McQuillan, J. M., "Why Go to Electronic Mail?" *Computerworld*, September 28, 1981, p. 9.

McWilliams, P., "An Introduction to Word Processing," *Popular Computing*, February 1982, pp. 17–30.

Mertes, L. H., "Doing Your Office Over—Electronically," *Harvard Business Review*, March–April 1981, pp. 127–135.

Miller, H., "Teleconferencing," *Computerworld*, September 28, 1981, p. 29.

Morgan, H. L., "The Future of the Office of the Future," *Office Automation Conference*, AFIPS, 1981, pp. 27–31.

Nolan, R. E., R. T. Young, and B. C. DiSylvester, *Improving Productivity Through Advanced Office Controls*. New York: AMACOM, 1980.

Norman, C., *Microelectronics at Work: Productivity and Jobs in the World Economy*. Washington, DC: Worldwide Institute, Paper 39, 1980.

Nussbaum, K., "Women Office Workers in a Race Against Time as Automation Impacts the Work Place," *Computerworld*, September 28, 1981, p. 40.

Nutt, G. J., and P. A. Ricci, "Quinault: An Office Modeling System," *Computer*, May 1981, pp. 41–57.

Paller, A., "Firms Finding Graphics Essential to Office," *Computerworld*, September 28, 1981, p. 43.

Panko, R. R., "Integration in Office Automation: Are We Putting the Cart Ahead of the Horse?" *Computerworld*, September 14, 1981, pp. 17–24.

Parkinson, C. N., *Parkinson's Law and Other Studies in Administration*. Boston, MA: Houghton Mifflin, 1957.

Planning for Office Systems (Course Notes), New York: IBM Systems Science Institute, 1981.

Priest, S. L., and V. J. O'Sullivan, "Who Should Manage Area of Word Processing?" *Computerworld*, September 28, 1981, p. 64.

"Principes de la Commutation par Paquet," excerpts from *Output*, September 1980–May 1981, p. 8.

Provisional Recommendations X.3, X.25, X.28, and X.29 on Packet-Switched Data Transmission Services. Geneva: The Consultative Committee on International Telephone and Telegraph, 1978.

Rhodes, J., "Office Automation: At Which Stage Are You?" *Computerworld*, September 28, 1981, p. 17.

Rosenbaum, R. M., "Standard Oil Unearths Plan for Office Success," *Computerworld, September 28, 1981, p. 65.*

Saffady, W., *The Automated Office: An Introduction to the Technology*. Silver Spring, MD: National Micrographics Association, 1981.

Sanders, D. H., *Computers in Society*. New York: McGraw-Hill, 1981.

Schumacher, E. F., *Small Is Beautiful: Economics as if People Mattered*. New York: Harper & Row, 1973.

Simon, H. A., "Prometheus or Pandora: The Influence of Automation on Society," *Computer*, November 1981, pp. 69–75.

Skees, W. D., *Computer Software for Data Communications*, Belmont, CA: Lifetime Learning Publications, 1981.

Slonim, J., L. J. MacRae, W. E. Mennie, and N. Diamond, "NDX-100: An Electronic Filing Machine for the Office of the Future," *Computer*, May 1981, pp. 24–36.

Sprowls, R. C., *Management Data Bases*. Santa Barbara, CA: Wiley/Hamilton, 1976.

Steiner, G. A., *Strategic Planning: What Every Manager Must Know*. New York: The Free Press, 1979.

Synott, W., "Boston Bank Takes Office Automation Highway via Well-Traveled Word Processing Route," *Computerworld*, September 28, 1981, p. 12.

"Telecommuting: Not All Roses," *Computerworld*, November 30, 1981, p. 22.

Thurber, K. J., *Tutorial: Office Automation Systems*. Los Alamitos, CA: IEEE Computer Society Press, 1980.

Toffler, A., *The Third Wave*. New York: Morrow, 1980.

Uhlig, R. P., D. J. Farber, and J. H. Bair, *The Office of the Future: Communication and Computers*. New York: North-Holland, 1979.

Ulrich, W. E., "Introduction to Electronic Mail," *National Computer Conference*, AFIPS, 1980, pp. 81–84.

Ulrich, W. E., "Office of the Future Requires Careful Planning: Examine Office Culture, System's Impact First," *Computerworld*, September 28, 1981, p. 36.

Views, H., "The Human Network," *Computerworld Extra*, September 1, 1981, p. 67.

Wilk, E., "Survey Sees Integrated Information Systems an Industry Reality for Users by Mid-80s," *Computerworld*, September 28, 1981, p. 58.

Wohl, A. D., "Office of Tomorrow: What to Expect," *Computerworld*, September 28, 1981, p. 5.

Wohl, A. D., "Tactics for Getting Started in Office Automation," *Office Automation Conference*, AFIPS, 1981, pp. 153–154.

Zarrella, J., *Word Processing and Text Editing*, Suisun City, CA: Microcomputer Applications, 1980.

Zassenhaus, H., "Office Technology: Tinker Toy or Tool?" *Computerworld*, October 12, 1981, pp. 17–22.

Zientara, M., "Companies Experiment with Telecommuting," *Computerworld*, November 30, 1981, p. 23.

Zisman, M. D., "Office Automation: Revolution or Evolution?" *Sloan Management Review*, Spring 1978, pp. 1–16.

Zloof, M. M., "QBE/OBE: A Language for Office and Business Automation," *Computer*, May 1981, pp. 13–22.

Glossary

absorption cost system   A cost-accounting system in which, for purposes of determining costs, all indirect manufacturing costs applicable to a period of time are absorbed by the products worked on in that period. Product costs therefore represent the sum of (1) costs directly assignable to the product, and (2) a proportionate allocation of all indirect manufacturing costs applicable to the period of time within which the product was made.

In a standard cost system, on the other hand, overhead costs are predetermined or are absorbed by the use of standard or normal overhead rates based on a standard or normal level of operations and at normal costs. This procedure results in variances due to overabsorption of overhead during periods of above-normal activity and unabsorbed overhead during periods of subnormal activity.

access time   The interval between the time data are called for or requested to be stored and the time delivery or storage is completed.

acoustic coupler   A data communications device that converts electrical data signals to and/or from tones for transmission over a telephone line, using a conventional telephone headset.

action officer   Army term for word originator.

activator development   A type of developing process employed on diazo duplicators. Both the diazo compound and coupler are contained in the coating of the copy material. An organic amine solution, applied via pressure roller, is used as a developing agent to release the couplers in the copy to form an image. The system is odorless, and is sometimes referred to as "pressure diazo" by AM/Bruning. See *diazo process.*

active document   A document requiring original thought. Research, organization, proofreading, and revision are generally required. Error-free copy is often needed.

activity list   The list of a department's main operations; defined by the department's head or supervisor.

address   A locator assigned to a specific position of keyed material on a recording medium.

adjust   A formatting procedure which allows line endings to be changed to conform with new margin settings.

administrative secretary   A secretarial specialist who handles such nontyping activities as the telephone, filing, reservations, mail, etc.

administrative study   A study that identifies and measures nontyping activities, including the levels and adequacy of support.

administrative supervisor   A supervisor of administrative secretaries.

administrative support   An office personnel concept whereby nontyping functions such as filing, telephoning, scheduling, and reservations are performed by a specialized secretarial team.

administrative support center   A physical location within an office where administrative support personnel function.

administrative terminal management system   An advanced form of ATS.

AGC   Automatic Gain Control. See *gain control.*

AGR   Annual Growth Rate.

ALC   Automatic Level Control.

algorithm   A group of programming routines which will cause the system to perform such processes as hyphenation, character spacing for justification, etc.

allowance (budget allowance)   The amount of expenditure designated for a specific purpose by the budget.

allowances   Nonproductive time computed into a standard to allow for coffee breaks, fatigue, rest time, and activity reporting.

alphabetic commands   Instructions to a word processor implemented by a code or control key plus alphanumeric key or keys. Such dual-function keys may be marked on the key cap, or the operator may be required to memorize most or all of the alphanumeric command set.

alternate sheet-feed   The ability of some offset duplicators to feed paper on alternating machine cycles (every other revolution of the plate cylinder) to allow heavier than normal plate-to-blanket inking for printing large solid areas.

AM   Amplitude Modulation. Modulation of the amplitude of a radio carrier wave in accordance with the corresponding amplitude of the audio or other signal.

ammonia development   The most common type of developing process employed in diazo duplicators. The coating of the copy material contains both the diazo compound and couplers. Ammonia vapor is used as a developing agent to release the couplers in the copy coating to form an image. Ammonia vapors may be derived from an aqueous ammonia supply (a 26% ammonia solution) or an anhydrous ammonia supply (vapor alone, contained in pressurized tanks). See *diazo process.*

amortization   The write-off period for purchasing equipment, most commonly five years.

analog   A term used in contrast to digital, relating to representation by means of continuously variable physical conditions. In facsimile, the term "analog" refers to the way in which a fax unit converts the optical data derived from scanning the original into electrical signals. Generally associated with slower facsimile units, the typical analog transmitter scans every inch of an original: the characters, spaces between characters, and the margins. Each picture element of the original is represented by an analogous electrical signal. A combination of these analog signals creates a continuous electrical waveform (current) which, when used to drive a printer, reproduces a series of picture elements which closely resemble the original.

announcement unit   Answers incoming phone calls and gives a prerecorded message. At end of message, unit usually switches audio from outgoing to incoming to permit callers to leave recorded messages.

anyhdrous ammonia development   See *ammonia development; diazo process.*

aperture   The adjustable lens opening of a camera or camera/platemaker. The size of a particular opening is designated by an f-stop number (e.g., f11, f16, f22, etc.); the larger the number, the smaller the amount of light allowed to strike the photographic material.

applicability, area of (or zone of)   Refers to an expressed cost-volume-profit relationship, such as a breakeven chart or a cost curve. The area of applicability is the range of activity within which the given relationships are realistic and express attainable costs or profits.

applications   Refers to a type of document processed for a particular purpose or in a special way.

applications software   Software usually provided by the vendor which allows the user to perform data processing or other tasks with the word processing system without having to write such software. Systems which employ such software are generally multipurpose, and may or may not provide a user programming capability.

applique   See *baseplate.*

apprenticeship   A system of training for the skilled trades that encompasses both formal classroom training and on-the-job training.

APS   Alphanumeric Photocomposer System.

aqueous ammonia development   See *ammonia development; diazo process.*

archive   A procedure for transferring information from an on-line storage diskette or memory area to an off-line storage medium.

arithmetic capability   The ability of a word processing system to be used as a calculator or adding machine. Some of the more sophisticated systems have the ability to do arithmetic tasks; a smaller number can do such tasks as part of word processing routines, with totals embedded in text, etc.

ARO   After Receipt of Order.

ASCII   American (National) Standard Code for Information Interchange. A seven-bit-plus-parity code established by the American National Standards Institute to achieve compatibility between data services. ASCII consists of 96 displayed characters and 32 nondisplayed control characters.

A-size   See *engineering size.*

ASK   American Simplified Keyboard. See *Dvorak-Type Keyboard* for definition.

ASR   Automatic Send/Receive. A communicating device that operates in an automatic send/receive mode of transmission.

assembler   A computer program that converts a higher-level (English-like) programming language into machine-readable instructions.

asynchronous   A mode of data transmission in which data are transmitted a character at a time preceded by a start bit and followed by a stop bit in order to ensure correct receipt. Commonly used for interactive communications.

ATMS   Advanced Text Management System. An advanced form of ATS.

ATS   Advanced Text System. An IBM software package which permits a computer to be used for word processing with text extended and retrieved via remote interactive computer terminals.

attendant phone   Allows dictator to communicate with attendant at remote recorder location in a central dictation system.

audiovisual   Literally "sound-sight." Refers to teaching or communications aids that use the senses of hearing and vision to convey their meanings more effectively. Some media examples are: tape-recorded sales demonstration; color slide presentation, with or without sound; blackboard; flip chart; movie film; and video (TV).

auditron   The Xerox trade name for an independently manufactured plug-in type copier monitor device. See *copier monitor device; plug-in counter.*

author's alteration (AA)   Any revision to a document by the originator.

automatic answer   A feature by which a communicating word processor may receive text without an operator in attendance.

automatic call   A communications feature that allows a transmission control unit to automatically establish a connection with one or more message recipients.

automatic camera/platemaker   A device for making offset press plates. It incorporates a camera or vacuum frame exposure unit, and a plate processor in one on-line system. Various types of automatic systems are available, including those which produce diffusion transfer, electrostatic, silver-halide, and wipe-on plates. See *electrostatic plate; transfer plate; wipe-on plate.*

automatic carrier return   Automatic performance of a carrier return when the last word which will fit onto a line is typed. The system generally has a buffer to hold the word currently being typed until it judges whether to place the word on the current line, or to wrap it onto the next line. Systems which automatically perform carrier returns are speedier on input, since the operator may type text at a uniform speed without pausing at each line end to perform a return.

automatic centering   The ability to automatically center a word or text segment between margins or a designated point. Function implementations may be by a single keystroke or multiple keystrokes for the centering of previously typed text or text as it is typed. Some systems have the capability to center material between tab settings.

automatic decimal tab   Automatic alignment of columns of decimal figures on the decimal point. The typist can type numbers without regard for alignment, with the system performing the aligning chore.

automatic dial   A communications feature whereby the calling unit has the ability to automatically establish a connection with one or more message recipients.

automatic document feeder   A device employed on a copier or automatic duplicating system for the continuous feeding of stacked single-sheet originals to the exposure platen. Sometimes known as a stack feeder. A special document feeder configuration will allow the feeding of continuous, unbursted computer printout forms. See *computer forms feeder; recirculating feeder; single-sheet document feeder.*

automatic duplicating system   A high-speed reproduction device employing electrostatic, direct lithographic, or offset techniques to produce multiple copies at a rate greater than 100 per minute. General machine operations, other than setup and shutdown, are performed automatically, as with a copier. Other equipment features may include automatic document feeding, image reduction, and on-line sorting of copies. See *direct lithographic process; electrostatic process; offset process.*

automatic etching   The ability of some offset duplicators to perform plate etching without operator intervention. See *etch.*

automatic file select   The ability of the system to make selections from data files, based on the characters which appear in a certain data field. For instance, using a zip code field, the system can select all the addresses with a 19101 zip code for one letter, and type a different letter to those in zip code 19104, etc.

automatic file sort   The ability of the system to arrange data in alphabetical, numerical, or other order. This feature is important on systems that sort or manipulate address lists; changes then need not be performed in alphabetical order.

automatic footnote tie-in   The ability of the system to tie footnotes to appropriate text segment. If a text segment is moved to another page or document, the footnote will travel with it. Generally, systems which are sufficiently sophisticated to offer footnote tie-in also automatically handle the division of space between the main text and footnotes.

automatic forward reset   Feature of central telephone dictation recorders in which the recorder automatically continues in playback mode to end of recording even after the reviewer has disconnected.

automatic headers/footers   The ability to place header/footer text at the top or bottom of each page of a multipage document. The operator specifies the text once, and the header/footer (usually document title, company name, or confidentiality requirements) is automatically added during printout. Changes may be made to the main document text without affecting the headers and footers.

automatic input underline   System allows the operator to indicate the beginning and end of an underline by a code, rather than backspacing and underlining on a character-by-character basis. Some systems allow the operator to choose between underlining words only, or underlining spaces and words. A few use a code that will underline the previous or next word, line, or other text segment.

automatic letter writing   The ability of a word processing system to merge a name and address list with broilerplate text to produce repetitive, personalized correspondence.

automatic line spacing   Different line spacings (single, double, triple, etc.) are permitted without performing physical setting changes on the printer. This enables the typist to input text with combinations of spacing without stopping and resetting the printer during playback printout.

automatic margin adjust   The ability to perform margin changes by a single command, automatically changing line endings without further intervention. During margin adjust procedures, temporary hyphens are normally dropped (unless they occur at the end of a line), and the system may give the operator an opportunity to make new hyphenation decisions to afford a "tight" line.

automatic measured review   A setting on the transcriber's foot-pedal for determining the number of words to be repeated upon depression of the foot-pedal.

automatic page numbering   The ability of a word processing system to automatically generate page numbers within a document. When text is rearranged and page numbers change, the system can generate a new set of correct page numbers. This feature enables the operator to input text without regard to final page endings. The system will create pages of the desired length, and number them appropriately.

automatic pagination   The ability to take a multipage document and divide it into pages of a specified length (in numbers of lines). Often, such ability is joined with the capability to automatically generate page numbers.

automatic repagination   An automatic routine to change page endings if text is inserted or deleted within a document, or if a new page length is desired. Text will be removed from or added to pages, as required, to maintain page length.

automatic repeat key   A "live" typewriter key (such as the underscore), which continues to operate as long as the key is depressed.

automatic reverse   The ability of some recorders to reverse at the end of a tape and to play back without having to change media.

automatic selector   An electronic switching system that directs dictation to the first available (open) recorder.

automatic send   A communications capability whereby the communicating system has the ability to automatically send out a message in an unattended mode.

automatic separation   See *separation.*

automatic typewriter   The simplest form of word processor. Used for straight, repetitive output with little or no text editing.

automatic widow adjust   The ability of a word processor to automatically prevent the first line of a paragraph, or a title or heading, from being the last line on a page. It may also prevent a paragraph's last line from being the first line on a new page. Such a feature is especially desirable if the system paginates or repaginates automatically.

automatic word recall   An adjustable feature of some transcriber units, whereby each time the foot pedal or hand control is depressed, a measured portion of the previous dictation is replayed.

automatic word wraparound   The ability of a word processor to automatically place a word which does not fit onto the line being typed onto the next line. Frequently combined with the *automatic carrier return* feature. Also used to denote systems which can wrap words during margin adjust procedures.

automation   A system of production in which work in process is transferred from one operation to another without human intervention.

autonomous work groups   A practice in which work groups are given relatively high responsibility for determining production methods and the jobs and tasks to be carried out by group members. It combines elements of job enrichment, participation, and often group incentives.

average letter   Approximately 92–115 words or 18–23 lines of 12-pitch typing. As the standards are hard to define, this method of measurement is generally unsuitable.

azimuth loss   Signal loss due to misalignment between playback head gap and the signal recorded on tape.

background processing   The automatic execution of a print function or, on some systems, a sort function, simultaneous to the keyboarding or editing of another document.

backup   A means of protecting valuable company information. May take the form of (1) duplicating tapes or disks on which data is stored; (2) providing a system with an alternate power source to protect data in volatile memory in the event of a power failure; (3) providing a redundant system.

balance-sheet budget   A pro forma balance sheet showing how each account will appear at the end of a budgeted period, or at any interim points, if the entire program of operations, including all incomes and expenses, the handling of assets and liabilities, and the capital budget, is realized.

band printer   Also known as a belt printer, an impact device which employs a metal band imprinted with characters which rotates horizontally past the paper. Impressions are made through the firing of hammers against the paper, ribbon, and belt/band.

bandwidth   The range of frequencies that can pass through a circuit. It is a measure of the rate at which information can be passed through the circuit.

baseplate   An interface device which connects to an ordinary typewriter, converting it (with an attached media console) into a low-level word processor.

BASIC   A higher-level, English-like computer programming language.

basis weight   The weight, in pounds, of 500 sheets (one ream) of a particular grade of paper cut to a standard size.

batch, batching, batch processing   A computer procedure where similar tasks are grouped and performed sequentially to increase productivity and simplify operations.

batch control   A control device that metes out predetermined quantities of work to an employee at regular intervals of time.

baud   In communications, a unit of transmission speed; generally a baud equals a bit (of data) per second.

baudot code   A data transmission code in which five bits represent one character.

BCD   Binary-Coded Decimal representation. A system of representing decimal numbers. Each decimal digit is represented by a combination of four binary bits.

beating the shift   A typewriter action whereby a very fast, or erratic, typist may cause a malprinting of a character following or preceding a shift.

belt printer   See *band printer*.

bias   A constant high-frequency tone added to the audio signal to overcome nonlinearity of magnetic media.

bidirectional printing   With bidirectional or bustrophedon printing, the system prints line 1 from left to right and line 2 from right to left, saving time by avoiding unnecessary carriage (or element) movement. A few systems which employ bidirectional printing will also check for the closest margin before deciding to print a line left to right or right to left; this can save a few seconds in printing a segment immediately following a very short line.

binary code   See *BCD*.

binder   A device that uses a method other than stapling to bind multiple-page copy sets. Methods employed include flat comb, spiral comb, cold glue, hot melt/glue, and thermal tape. See *flat comb binder; hot melt/glue binder; spiral comb binder; thermal tape binder.*

bisynchronous   A set of operating procedures originated by IBM for the synchronous transmission of binary coded data.

bit   The commonly used abbreviation for binary digit. A bit is the smallest unit of information recognized by a computer, and is a unit of information corresponding to a choice between two alternatives (such as one and zero).

black box   Generally defines any electronic module that does something to a signal flow, like automatically translating from one message protocol to another protocol. A system designer is generally not concerned with the insides of the box, only with its input and output.

blanket cylinder   In offset duplicating, a rubber-covered cylinder which transfers the inked image from the plate cylinder to the printing paper. See *offset process.*

bleeding   A term referring to the "splashing" or "bleeding" of carbon material from a carbon ribbon onto the paper or into the typewriter mechanism; a condition when red ink from a red and black fabric ribbon "bleeds" into worn-out black portion of ribbon.

block move/copy   The ability to designate a block of text (generally with some maximum number of characters and related to buffer size) and to move it within the document or to another document. Most systems which can access and move blocks of text can also copy blocks to another storage location (for editing) without erasing the original text.

boilerplate   Stored paragraphs which may be combined to create a new document. Variable information, either prerecorded or keyboarded, may be combined with such boilerplate in most systems.

bond   A common, relatively high-grade paper stock employed in the office for letters, business forms, and copying. Bond usually has a rag content ranging from 25% to 100%.

breakage   The difference between equivalent manpower and actual manpower.

broadband   See *wideband.*

brush   A device used in plain-paper copiers to remove excess toner from the photoconductor following image transfer. The brush wipes toner residue from the drum and deposits it in a disposable filter bag.

BSC   Binary Synchronous Communication. An IBM designation referring to a specific communications procedure using synchronous data transmission.

B-size   See *engineering size.*

bubble memory   A new nonvolatile storage technique which uses magnetic fields to create regions of magnetization; a pulsed field breaks the regions into isolated bubbles, free to move along the surface of the crystal sheet which contains the regions. The presence or absence of a bubble represents digital (bit, not bit) information. External electromagnetic fields manipulate the bubbles (information) past "read/write" locations within the memory. This is analogous to the motion of a read/write head in disk storage, or magnetic tape motion past read/write heads in magnetic tape storage. Because bubble devices are very tiny and are nonvolatile (information is not lost when current is interrupted), this variety of storage is likely to find wide application in terminals, word processors, and other office devices, particularly when volume production decreases cost.

buffer   A high-speed area of storage that is temporarily reserved for use in performing the input/output operation, into which data are read or from which data are written.

buffer configuration   The way in which the system buffer may be utilized, especially whether it may be split (and where, if restricted) to allow text (e.g., a letter) to be merged with data (e.g., names and addresses).

buffer size   The number of characters of text and command codes a system can manipulate at one time.

buffer storage   A device into which information is assembled and is stored pending transfer.

bug   A program defect or error.

bundled   See *unbundled*.

burden   See *overhead*.

bustrophedon printing   See *bidirectional printing*.

bypass feed   A convenience feature of some roll-fed copiers that enables the user to manually feed single sheets of pre-cut copy paper in sizes other than that of the loaded paper roll. A bypass feed may also be employed to generate offset masters, transparencies, and label material.

byte   A sequence of eight adjacent binary digits that are operated upon as a unit and that constitute the smallest addressable unit in a computer or word processing system.

cadmium sulfide   A possible coating for a photoconductor, commonly used on drums for plain-paper liquid toner systems. See *photoconductor; plain-paper process*.

camera/platemaker   See *automatic camera/platemaker*.

camera-ready   Copy ready to be photographed for reproduction.

capacity   A term describing the number of bits encoded onto a magnetic disk or diskette for the storage of information. Capacity is governed by the number of serial bits per inch recorded and the number of tracks on the media. Generally describes the amount of text which may be stored on one unit of magnetic media, expressed in number of characters or pages. See *double density*.

capstan   The driven shaft in a recorder—usually the motor shaft—which revolves against the tape and pulls it through the machine during recording and playback.

carbon sets   File and tissue copies, etc., manufactured with carbon paper attached.

carbonless or carbon-coated business forms   Business forms that permit impressions from copy to copy without the use of carbon. The image is made when special coatings on the back of one sheet and the face of the following sheet are brought together under pressure.

card or key station counter   A device used to monitor copier usage. A terminal is mounted on a copier that contains a number of stations (locks) and corresponding digital counters. Depending on the system employed, a coded card or key is used to unlock a station and activate the copier for usage; all copies made by the user of the card/key are tallied in the proper terminal meter. A periodic reading of terminal counter figures is used to charge back departments for copier usage. See *copier monitor device*.

cardioid microphone   A directional microphone, having a heart-shaped pattern, rejecting sounds from the rear.

carriage paper width   Limits the maximum paper width. Maximum writing line is usually two inches less than physical paper width.

carrier return   See *automatic carrier return*.

cartridge   See *magnetic media*.

cassette   See *magnetic media*.

cassette-feed   A method employed on some copiers for the feeding of sheet stock. Various paper sizes are loaded into different plastic cassette trays. A change in copy size simply involves the interchanging of cassettes. A dual-cassette-fed copier provides two console-selectable paper sizes available at all times. A small number of copiers employ disposable preloaded cardboard cassettes.

CCITT   The Consultative Committee on International Telephone and Telegraph is a United Nations group currently setting up worldwide communications standards.

CdS   See *cadmium sulfide*.

CE   Customer Engineer.

centering   See *automatic centering*.

central dictation system   A class of dictation equipment designed for local and/or remote point-of-dictation, multi-originator facilities where all dictation is pooled in a central recorder or "tank." The dictator accesses the system via a personal microphone or handset-interfaced network. One or more typist-transcriptionists continually monitor the tank and transcribe dictated correspondence. Central systems may be configured to handle from two to an unlimited number of dictators.

centralization   An office systems design in which all equipment is centrally located in one area. This approach allows for more complete control over the cost and operation of equipment. See *decentralization*.

centralized loops   Several trunks housing continuous-loop magnetic tapes that receive dictation from several stations.

central office line   A direct-dial telephone line.

central recorder   A device placed in a central location to receive dictation from several stations.

centrex   A system that provides each telephone station with its own number; installed in companies with extremely heavy telephone traffic. Calls do not go through a switchboard.

chad The small pieces of paper tape or punch card removed when a hole is punched.

chain delivery A type of paper delivery system customarily found on higher-volume offest duplicators. Gripper bars, mounted on two recirculating chains, grip the foremost edge of the printed sheet and guide it to the delivery stacker tray. See *gripper margin; grippers.*

chain printer An impact printing technique in which a set of character slugs moves horizontally past a set of hammers. As the character slugs pass in sequence, hammers are fired to imprint each required character through a ribbon onto paper. The character slugs are connected and pull each other around a track. More than one set of characters may compose a chain. Another method of using character slugs is to mount them in a track which moves. This latter technique is called the train process.

character generation The technique used to generate characters on the display screen. See *dot matrix.*

character set Total number of different characters displayable, including alphabetics, numerics, and special symbols. Alphabets may be shown as upper case only, or upper and lower case.

character size control The ability of a display screen to allow the operator to view a full page of text at regular character size or one-half page at double (vertical) size. A few systems also allow the operator to view a double horizontal page (for wide-page work), such as tabular reports, at half-character size.

character spacing display The ability of a word processor to show characters in different pitch and/or proportional spacing on the display.

charge In an electrostatic output device, the electrical charge of the surface of the photoconductor.

charge system The use of special log cards for recording removal of file folders from files.

chip A microprocessor that is a complete computer on a single chip of silicon. No larger than an inch square, a chip contains all of the essential elements of a central processor, including the control logic, instruction decoding, and processing circuitry. To be useful, the microprocessor chip or chips are combined with memory and I/O integrated circuit chips to form a "microcomputer," a machine almost as powerful as a minicomputer. These chips usually fill no more than a single printed circuit board.

chute delivery A type of delivery system found on smaller offset duplicators. As printed sheets are ejected from the impression cylinder mechanism, they are delivered down a chute to a receiving tray.

clamp Braces located on the plate cylinder of an offset duplicator. They are employed to fasten the leading and trailing edges of a plate to the surface of the cylinder.

clean corotron See *corotron.*

clerk An employee responsible for correspondence records or accounts or for general office work.

clipping A condition in dictation where the first part of a word is not recorded as the recording mechanism engages.

coaching The process of developing a subordinate through counseling, trial assignments, review, and analysis of performance. Usually refers to the development of a subordinate supervisor or manager. Normally implies more than usual day-to-day supervision, requiring that there be a planned program of development, with specified objectives and means of measuring progress toward their accomplishment.

coated paper (1) Copy paper that has been treated with a charge-sensitive substance, allowing it to be used in the electrostatic copying process. The coating of the paper is usually zinc oxide. (2) A grade of paper used in offset duplicating. It consists of a mixture of clay and other materials applied to the surface of plain-paper stock. See *coated-paper process; electrostatic process.*

coated-paper copier See *coated-paper process; electrostatic process.*

coated-paper process An electrostatic copying process in which the image of the original is projected directly onto the coated, charge-sensitive surface of the copy paper. The coated copy paper acts as a photoconductor. See *coated paper; electrostatic process; photoconductor.*

coaxial cable A cable consisting of one conductor, usually a small copper tube or wire, within and surrounded by a shield made of a separate electrically insulated wire.

COBOL A higher-level, English-like computer programming language.

code conversion The translation from one type of code to another.

code set A specific set of symbols and rules used to represent information.

cold start The activation of a copier's power after a period of nonuse, such as in the morning of a working day. At this stage the machine may or may not require a warm-up period, depending on the type of copier.

cold type Typesetting normally produced by a direct impression typewriter mechanism.

collator A device that sorts multiple copies into groups or sequenced copy sets. The term "collator" usually refers to a low-volume application in which the unit is operated off-line from a copier or duplicator. Collators are generally table-top devices, and may be manually or semiautomatically operated. See *sorter.*

column move/delete   The ability of a word processor to manipulate characters vertically within a column. This feature is important for tabular work, since a column can be moved or deleted with a minimum number of commands. In less sophisticated systems, columns must be moved or deleted a line segment at a time in multiple steps.

column wrap   The ability of a word processing system to automatically readjust and rewrap text among columns to conform to a new format or to a specified format after an insertion, deletion, or edit operation has been performed.

COM   Computer Output Microfilm. Normal printed output of a computer reduced to one of several available microforms by a special output device that takes the place of the normal print output device. The COM unit allows high-quality output at speeds of 5,000 or more lines per minute. Computer magnetic media files are fed directly into a recording device for rapid preparation of the data, and output to extensively reduced film images. Microform viewers are used at strategic locations for rapid dissemination of information. The system has the advantages of rapid access to vital business data, and reduction of storage space.

combination unit   A desktop unit on which a dictator records information and from which a secretary transcribes the dictation.

command   A signal or group of signals which causes a word processor to execute an operation or series of operations.

common carrier   A government-regulated private company that furnishes the general public with telecommunications service facilities; for example, a telephone or telegraph company.

communications network   The connecting of geographically separated communicating devices via transmission lines. A local network connects users within a limited geographic area while a remote network connects widely dispersed locations.

compatibility   A characteristic of word or data processing equipment which permits one machine to accept and process data prepared by another machine without conversion or code modification.

compatible model   A device marketed by a manufacturer which is also private labeled and sold by other companies in the same competitive field.

compiler   Software that translates program instructions written in a high-level language such as BASIC, FORTRAN, COBOL, or PL/I into object code (machine level language) for execution by the system. The presence of a compiler on a word processing system indicates that the system employs a multipurpose computer and that the manufacturer allows or encourages user programming.

computer, digital   A computer that functions by interpreting discrete electronic impulses.

computer forms feeder   A type of automatic document feeder designed to handle continuous-form computer printouts. The material is fed to the exposure platen of a copying device, where reproductions are made with a minimum of manual paper handling. See *automatic document feeder*.

computer programming   Providing a set of instructions for a computer.

concentrate   See *toner concentrate*.

concurrency   The ability of a communicating word processor to send and/or receive messages in background, simultaneous with the entry or editing of text. This feature increases the total throughput of a system.

conditional sale   A type of *financial lease* under which the customer acquires title to the equipment. Part of the monthly rental will be applied toward the total price of the equipment, with a nominal end-of-lease purchase price being employed to conclude the arrangement. For tax purposes, the conditional sale is treated as a purchase. Conditional sales may also be termed rent-purchase or lease-purchase plans, and are available from manufacturers and third-party lessors.

configuration   The components which make up a word processing system. Most systems include a keyboard for text entry, a form of magnetic storage (such as cards, cassettes, or diskettes), and a printer for output. A number of systems also include a video display (from less than one line to a full legal-size page) to view text entry, editing, and system status. Some systems may include a minicomputer, and a number of special peripherals (OCR, line printer, computer tape, etc.).

connect time   Elapsed time during which a terminal or communicating word processor is connected to, and functioning as a station of, a computer.

console   The unit housing the record/playback mechanism on some text-editing typewriters.

contact print   A same-size print made from either a film negative or a positive brought in direct contact with sensitized paper, film, or plate material. The material is secured by a vacuum frame, exposed to a light source, then processed and dried. See *exposure*.

content search   The ability of a word processing system to search through text to match a group of characters.

continuous form   A supply of paper made up of numerous individual sheets separated by perforations and folded to form a pack. Sprocket holes are punched in the margins to permit automatic feed through a printer.

continuous-loop recorder   See *endless-loop recorder.*

continuous tone   A photographic image containing gradual tones from black to white; must be screened for printing. See *halftone; halftone screen.*

contrast   The density of blacks and whites in a copy image. See *contrast control; exposure.*

contrast control   A console control on a copier used to adjust the lightness or darkness of copy images. An adjustment of the control usually regulates the amount of toner being applied to the copies, although in a few cases the contrast control may alter the exposure time to effect the same results. See *exposure.*

control   A warning system (logs, numerical controls, etc.) that forecasts potential bottlenecks and affords sufficient clues for correcting any problems, errors, or fall-downs.

control character   A coded character which does not print but initiates some kind of machine function such as a carrier return.

control-character printout   The ability of the system to provide a printout showing all normally concealed control characters (required carriage returns, indent commands, etc.); line numbers may also be shown on such a draft printout.

control-code display   The ability of a word processor to display instructions, commands, or codes on the video screen. In some systems the operator may choose between displaying text with codes, or text only (as it will appear on printout).

control counter   See *card or key station counter; copier monitor device.*

control graph   A technique for charting the level and distribution of control among various segments or levels of an organization.

controlling   The managerial function that keeps the unit progressing steadily toward the accomplishment of its objectives and fulfillment of its purpose. Involves a system of observation to determine if standards are being met and if objectives are being accomplished; some form of measurement to assess the deviation; and corrective action where needed.

conversational mode   Communication between a terminal and a computer in which each entry from the terminal elicits a response from the computer and vice versa.

converter   See *media converter.*

coordination   Ensuring that all efforts are bent toward a common objective and that there is no duplication of work that results in wasted effort. Includes resolution of differences of opinion.

copier control device   See *copier monitor device.*

copier-duplicator   A term applied to electrostatic copiers with speeds greater than 50 copies per minute.

copier monitor device   A device implemented on a copier to record all copies made and identify machine users. The typical system consists of an access terminal attached to a copier which may only be activated by designated keys, coded cards or keyed-in I.D. numbers. Such systems improve copier management by limiting machine access and providing chargeback data. See *card or key station counter; electronic event logger; plug-in counter.*

copy   Material worked from to prepare final or hard copy.

copyboard   A flat exposure area on which original material is positioned for reproduction by an overhead camera. The original is usually sandwiched between the alignment board and a glass plate. The term is generally associated with camera/platemakers. See *exposure platen.*

copy quantity selector   A console control of a copier that allows the operator to select the number of copies needed. Control types include dial, push button, thumbwheel and keypad configurations. Special features of some selectors include "M" (multiple) or "C" (continuous) settings for copy runs beyond the normal limit of the control, and auto-reset, which counts down the copies as they are made and returns the selector to the "1" copy setting.

copy revision   Includes console preparation, automatic playback when making corrections due to editorial changes, and actuating of various playback controls.

core   Usually employed to denote the storage size of computer (CPU) core memory.

corona   See *corotron.*

corotron   In an electrostatic printer, an electrostatic charging device, usually composed of a taut wire, which charges a photoconductor to facilitate imaging in the printing process. Three types of corotrons, or coronas, are generally employed in a photoconductor imaging system: *charge corotron,* which charges the surface of the drum positively, preparing it to attach negatively charged toner particles; *transfer corotron,* which carries a much higher positive charge than the drum and causes toner to be transferred to the surface of the copy paper; *clean corotron,* which emits a negative charge to neutralize the drum charge and prepare it for cleaning. See *electrostatic process; photoconductor.*

correction fluid   A white liquid coverup applied over an error.

correction paper   A chalk-coated strip of paper. The typist places the coated side against the error and retypes to cover the error.

correlation   A statistical term referring to the extent to which two variables or measures parallel each other, i.e., whether the scores which individuals or groups obtain on one measure (e.g., job satisfaction) are indicative of how high or low they perform in terms of some other variable (e.g., absenteeism). The correlation coefficient is an index of the closeness and direction of correlation, and ranges from +1.00 (perfect positive relationship), through 0.00 (no relationship), to −1.00 (perfect inverse relationship).

correspondence center   A secretarial group performing typing activities.

correspondence secretary   Word processing operator.

correspondence study   A study that identifies typing activities and measures the volume of typing in an organization.

cost-behavior pattern   The manner in which a cost fluctuates because of changes in the volume of operation, production, or sales.

cost curve   A graphic representation of the behavior pattern of a cost.

cost formula   An expression of the values and relationships pertaining to the constituent elements of a cost, from which a total cost may be computed.

cost standard   A computed normal cost with which costs actually incurred are compared. Cost standards are not incorporated in the formal books of account.

costs, standard   Unit costs set in advance of production by determination of the labor and material needed, plus a standard amount of overhead.

costs, variable   Costs that change with the level of company activity.

counter   A resettable digital indicator used to count the number of copies or impressions in a copy or duplicating run.

coupler   See *diazo process.*

CPC   Coated-Paper Copier. See *coated paper; coated-paper process.*

CP/M   Control Program for Microcomputer, an operating system developed by Digital Research.

CPM   Critical Path Method.   A management tool for planning and control; requires arrow-diagramming all activities in a given project, showing their relationships and estimates. From this can be identified the "critical path," the connected sequence of activities composing the longest duration path. (It is formed of activities that cannot take place completely simultaneously but must be at least partially completed before a related activity can be started.) CPM facilitates tight scheduling, visibly relates time schedules to cost, tightens control practices, and highlights effects of possible delays or obstacles. Differs from PERT in separating planning from scheduling and directly relating time and costs. See *critical path scheduling; PERT.*

CPM   Copies Per Minute, as in copying.

CPS   Certified Professional Secretary.

CPS   Characters Per Second. A unit of measure equal to the number of characters an output device is capable of printing in one second.

CPU   Central Processing Unit. The heart of a computer that controls the interpretation and execution of instructions.

crank-up time   Nonoperating time, not chargeable to any other function.

critical path scheduling   A method of scheduling work by means of diagrams that show which jobs must be completed before other jobs can be started. Jobs are indicated by arrows; hence the technique is sometimes called arrow diagramming. See *CPM*.

crossfooting   The addition, horizontally, of rows of numbers, with the sum placed at the last position on the line.

CRT   Cathode Ray Tube, a TV-screen-like device used to display text.

CRT flying-spot scanner   A type of scanning mechanism that uses a CRT-generated beam of light to raster-scan a stationary document mounted on a flat-bed platen. Through lens optics the reflected beam of light is focused onto a photomultiplier and converted into an electrical signal. High resolution is a primary characteristic of this method.

C-size   See *engineering size*.

CT/ST   Cassette Tape/Selectric Typewriter.

cueing   An audible signal that indicates end of dictation or special instructions from the dictator to the transcriptionist.

cumulative counter   A non-resettable copy meter employed by the copier manufacturer for usage/billing purposes. The meters are generally located inside the copier, and may register such things as total number of copies made, number of originals copied, and number of copies made within specific multiple-copy ranges (1 to 5 copies per original; 6 to 10 copies per original; 11+ per original, etc.).

cursor   A lighted indicator that marks the current working position on a display.

cursor positioning   Describes the motion of a cursor on a display. Most systems employ a series of arrow keys for up, down, left, and right movements. Some systems use a home key to position the cursor at the upper left corner of the screen. Some systems use a code key plus alphanumeric or function keys for cursor movement. A number of systems permit only horizontal cursor movement along a fixed line.

curved platen   An exposure board on a copier that is curved outward (convex) rather than flat. See *exposure platen*.

cybernetics   The study of control and communication in animals and machines.

cylinder printer   A type of impact printer which employs a complete character set embossed on a series of rings around a small cylinder. The cylinder is rotated and shifted up and down to position each character as needed, while a hammer strikes the cylinder and presses it against a ribbon which creates an image on the page.

DAA   Data Access Arrangement. A direct interface attachment that connects a data communications device to a telephone line. The DAA is a permanent attachment, and protects the public telephone network from a sudden surge of power or interference from a data communications device.

daisywheel   The print element for a daisywheel printer, such as those offered by Diablo and Qume. Printwheels are interchangeable, allowing the operator to select an appropriate font. See *element printer; impact printer.*

daisywheel printer   An interchangeable-element electronic impact printer, offering faster print speeds than a Selectric typewriter-printer. See *element printer; impact printer.*

dampening form rollers   See *form rollers.*

dampening solution   See *fountain solution.*

dampening system   The fountain reservoir and roller train which supply fountain solution to the plate cylinder of an offset press. See *fountain solution; fountain reservoir; roller train.*

data analysis   The evaluation and analysis of collected data.

data base   A nonredundant collection of interrelated data items processable by one or more applications. Nonredundant means that individual data elements appear only once (or at least less frequently than in normal file organizations) in the data base. Interrelated means that the files are constructed with an ordered and planned relationship that allows data elements to be tied together, even though they may not necessarily be in the same physical record. Processable by one or more applications means that data are shared and used by several different subsystems.

data base management system   A systematic approach to storing, updating and retrieving information stored as data items, usually in the form of records in a file, where many users access common data banks.

data collection   The gathering of data used to measure both administrative and correspondence activities within an organization.

data communications   The transmission and reception of encoded information over telecommunication lines.

data entry unit   A unit that transmits data into a computer, generally called a terminal.

data processing   The execution of a programmed sequence of operations upon data. A generic term for computing in busines situations and other applications with machines such as bookkeeping machines, digital computers, etc.

date stamp   A rubber stamp with dials which rotate to indicate the current date for stamping onto documents.

DDD   Direct Distance Dialing. The facility used for making long-distance telephone calls without the assistance of a telephone operator. DDD is frequently used to mean the switched telephone network.

DDS   Dataphone Digital Service. An all-digital transmission service offered by Bell Telephone. Through the direct connection of digital devices the need for special digital-to-analog modems is eliminated.

dead keys   Electric typewriter keys that do not automatically advance to the next character position when struck.

debug   Checking the logic of a software program to isolate and remove any mistakes.

decentralization   An office systems design in which reprographics equipment, primarily copiers, is located in departments throughout a company and used by casual operators. See *centralization*.

decimal tab   See *automatic decimal tab*.

dedicated copier or duplicator   A copier or duplicator that is employed within a company for a singular or special application, such as large-document copying, check copying, or computer-forms copying. Other types of copying (i.e. letters) should be considered low priority for this machine, and kept at a minimum.

dedicated operator   An employee whose primary job function is the operation of a specific piece of office equipment, such as a high-speed copier or duplicator. Dedicated operators are sometimes used in conjunction with a centralized reprographics system. See *centralization*.

dedicated recorder   Recorder devoted exclusively to one type of dictation or to specific individual(s).

default format statement   Some systems which employ format statements (information on margin settings, tabs, etc. internally stored for automatic implementation) have default format setting, with commonly used margin settings and sometimes a tab grid (usually every five positions). This default setting will be automatically implemented whenever the operator fails to designate a specific format statement.

degausser   A device used to erase magnetic tape without removing it from the reel.

delete capability   Indicates the grammatical segments (e.g., character, word, line, sentence, paragraph, page) by which text can be deleted from the storage media (and the display if appropriate).

delivery system   The mechanism of a copier or duplicator which transports completed copies from the imaging/impression section to the receiving tray or sorter. See *chain delivery; chute delivery*.

Delphi method   A technique for forecasting future developments, especially technological developments.

density   (1) The image contrast degree of blacks (and whites) of an original or copy. (2) The amount of toner on a copy surface. See *exposure; contrast control.*

departmental cost   The summation of the cost of operating a department (as distinguished from the costs of producing a unit of product).

desktop dictation unit   A type of dictation unit functioning in the classic one-to-one, dictator-to-secretary environment. Each originator has a personal desktop "dictator" (dictation machine) to record messages onto media; it is then physically forwarded to a transcriptionist for typing. Desktop units are also utilized as components of work-group units having remote dictate stations on several executives' desks and recorders usually located at the transcriptionist's desk.

developer   In the electrostatic copying process, a charged substance that acts as a carrier for the toner particles. Developers may be classified as dry (minute metallic or glass beads) or liquid (dispersant, solvent). See *electrostatic process; toner.*

diagnostic   Pertaining to the detection, discovery, and further isolation of an equipment malfunction or processing error.

diagnostic code   An alphanumeric or word display that signals a system condition such as a malfunction. The code is either self-explanatory or used to refer to further instructions that are explained in an operator guide.

diazo duplicator   A special-applications duplicator limited to the reproduction of one-sided translucent originals. Commonly applied to duplicate oversized documents such as engineering and architectural drawings, diazo-process machines are available with three basic types of development systems: activator, ammonia, and moisture. Also known as whiteprinters. See *activator development; ammonia development; diazo process; moisture development.*

diazo material   The reproduction medium employed on diazo duplicators. The material, which must be coated with diazonium compound, may be opaque paper, translucent paper, polyester film, acetate, or cloth. See *diazo duplicator; diazo process.*

diazo paper   See *diazo material.*

diazo process   A reproduction method employed on diazo duplicators. The underlying principle is that diazo compounds (diazonium salt) decompose under exposure to ultraviolet light and that a further reaction between the compound and a coupler forms a visible azo dye. In respect to reproduction equipment, the diazo compound and coupler are usually contained in the coating of the copy material. Exposure of the material to ultraviolet light create a latent copy image, and development in a basic environment (most commonly ammonia vapor) releases couplers which bring out the azo dye image. See *activator development; ammonia development; diazo material; moisture development.*

diazonium compound   See *diazo process.*

dictation speed   Median dictation speed of 60 words a minute.

dictation system   See *central dictation system; desktop dictation unit; portable dictation unit.*

dielectric process   A nonimpact printing technique in which specially treated paper consisting of a conductive base layer coated with a nonconductive thermoplastic material is used to hold an electric charge applied directly by a set of electrode styli. The electric charge corresponds to the latent image of the original. Following the charging step, the paper is imaged via a toner system similar to that of other electrostatic copying devices. This technique is sometimes called electrographic, and is currently employed on general-purpose nonimpact printers, facsimile devices, and some photocopiers. See *electrostatic process.*

diffusion transfer plate   See *transfer plate.*

digital   Pertains in general to information represented by a code consisting of a sequence of discrete elements. Also, a type of facsimile equipment which utilizes a digital transmitter to reduce the time required to send a document by eliminating redundant image data, i.e. the blank areas of an original. As a document is scanned, the unit employs run-length coding to assign brief on-and-off digital bit patterns to various types of picture elements and white spaces. Because the digital codes are more compact than the continuous stream of electrical signals associated with analog systems, the scanning, transmitting, and printing times are decreased.

direct-entry typesetter   The direct inputting of format and text material to a photocomposer via an integral keyboard arrangement, usually accompanied by some type of display. The term is synonymous with *direct-input typesetter.*

direct-impression master   A master that is made via typewriter, pencil, pen, or stylus for use on spirit, stencil, or offset duplicators. See *plate.*

direct-input typesetter   See *direct-entry typesetter.*

direct lithographic process   A type of impression system used on some duplicators and automatic duplicating systems. It is a two-cylinder system which images the paper between a plate cylinder and an impression cylinder. Plates are usually mounted on the plate cylinder automatically, and ejected after being used. Form rollers are employed to carry ink and water to the plate cylinder. See *form rollers; lithography; plate.*

direct lithography   See *direct lithographic process.*

direct reverse search   The ability of a system to search backward through a storage medium without having to first return all the way to the beginning of the medium and search forward.

directional option   The ability to transmit from one direction to another or both ways.

directory   See *index.*

discrete   Individually distinct.

discrete media   Individual magnetic tapes, belts, or disks that can be removed from a dictation unit.

discretionary hyphen   A semipermanent hyphen inserted by the operator in words that may require a hyphenation decision. Upon printout, the word processor can use any hyphen, or ignore it if no hyphen is required. Also called ghost or soft hyphen.

disk   See *magnetic media.*

diskette   See *flexible diskette (floppy) under magnetic media.*

dispersant   See *developer.*

display   A visual image, usually provided on a TV-type screen called a CRT.

display buffer memory   Size of the buffer holding characters displayed on the screen; total size may be larger than the number of characters actually displayed to allow a partial-page screen to scroll through a page or more of text.

display highlighting   The ability of the word processor to intensify or blink certain portions of the display screen—either the characters themselves or the screen area behind the characters—to emphasize a text segment designated for some special activity such as delete or move.

distortion   Any difference between the regular audio signal and the signal played back by the recording device.

distributed-logic word processing system   A multiterminal (station, keyboard, etc.) system which shares peripherals and sometimes storage, but where computing power (logic) is dispersed among individual stations or systems components. See *shared system.*

distribution   Delivery of a document to its final destination.

ditto   A reproduction process from purple-inked carbon masters.

document   See *copy.*

document assembly/merge   A word processing feature by which a system can assemble new documents from previously recorded text. Most systems can combine prerecorded text with keyboarded text.

Many systems can combine selections from prerecorded text to form a new document. Also describes the system's ability to join a document to such variable information as names and addresses to create a number of nearly identical documents.

document feeder See *automatic document feeder; computer forms feeder; single-sheet document feeder.*

Dolby system Circuitry used primarily in audio recording equipment to reduce the amount of noise, principally tape "hiss" introduced during recording.

dot matrix A method of display character generation in which each character is formed by a grid or matrix pattern of dots.

double density Term describing the storage of information on a floppy diskette such that the capacity is twice that of a standard diskette. This is accomplished by either doubling the number of tracks per inch, or doubling the serial bit density, or a combination of both.

double-sided diskette A type of diskette that utilizes both of its sides for the storage of information. A double-sized diskette can be loaded into a floppy-disk drive with a dual read/write head assembly; or used on a standard single-head drive, taken out, flipped, and reinserted for read/record operations on both sides. Also called a "flippy."

double voicing A condition most commonly found in earlier magnetic-belt recording systems. It describes an echo effect noticeable until the transcriber sound head is turned to the same place where the recording was made.

downtime The period of time during which a system is malfunctioning or not operating correctly because of mechanical or electric failure.

DP See *data processing.*

draft A rough and unedited manuscript.

dropout During playback, the brief loss of a recorded signal due to tape imperfection.

drum See *photoconductor drum.*

drum printer A type of printer that employs a rotating cylinder. A complete set of characters is embossed on the circumference of the drum for each print position. A set of hammers is used to strike the drum (through the paper and ribbon) and print the required character each time the drum rotates.

dry toner See *toner.*

dry-toner process An electrostatic copying method employing dry carbon toner particles suspended in a dry developer carrier. Dry toner may be applied via a photoconductor intermediate (plain-paper process) or directly to the copy paper (coated-paper process). See *electrostatic process; toner.*

D-size   See *engineering size.*

DSK   Dvorak Simplified Keyboard. A keyboard arrangement first patented in 1932 by Dr. August Dvorak of the University of Washington in Seattle. This keyboard arrangement claims a faster productivity over the conventional QWERTY keyboard arrangement.

dual column   The ability of a word processing system to allow text formatted in a single column block to be reformatted into two side-by-side columns. Dual-column capability also may implicitly refer to some degree of text entry/manipulation capability within each individual column without affecting the other column except in a readjust procedure.

dual density   See *double density.*

dual-media typewriter   Automatic typewriter with the capability of using two media—magnetic card and/or cassette.

dual pitch   Selection on one typewriter of pica or elite characters.

dual-spectrum process   A copying process requiring two separate types of paper. Light is used to create a latent image on an intermediate paper, and the intermediate is then in turn used to produce an image on a final copy paper by means of heat. The transfer procedure from original-to-intermediate to intermediate-to-final copy is performed manually by assembly from sets, or automatically by intermediate transfer within the machine. Dual-spectrum copiers are applicable to very low volume convenience copying.

dual track   A system using two tracks at the same time, such as a tape with two tracks—one for dictation and one for transcription.

duplex   In communications, the ability to send and receive information simultaneously. Sometimes referred to as *full duplex.* In printing/copying, a print/copy with images on both sides of the paper made by the electrostatic process. This is accomplished by the manual or automatic refeeding of one-sided copies through the exposure process, with the second image (second original) being fused to side two of the paper.

Dvorak-Type Keyboard   A simplified keyboard scientifically arranged to produce less finger movement, fatigue, and typing errors. See *DSK.*

EBCDIC   Extended Binary Coded Decimal Interchange Code. An eight-bit code originated by IBM that can accommodate 256 characters.

econometrics   A technique of making economic predictions by use of mathematics.

editing   Revising text with a word processor to update a document.

EDP   Electronic Data Processing. See *data processing.*

E-fax copier   See *coated-paper process.*

EIA interface   A standardized set of signal characteristics (time duration, voltage, and current) specified by the Electronics Industries Association for connection of terminals to modem units, and specific physical coupler dimensions specified by the Electronic Industries Association.

electrofax copier   See *coated-paper process.*

electrolytic process   A facsimile printing process employing a moist, electrolyte-impregnated roll-fed paper. The current (signal) is applied by feeding the paper between a stationary contact and a revolving drum with a helical contactor. The revolving of the drum causes the point of electrical contact between the blade and helical wire to move laterally across the paper, one line per revolution. As the electrical current passes through the paper it causes a dark · change of coloration at all points of contact.

electronic event logger   A device used to monitor copier usage. A terminal is mounted on a copier that monitors all copier usage and dispatches the data to a central recorder that stores the information on magnetic media. Some systems require copier users to activate the system with personal or departmental ID cards and/or keypad-entered number codes. The magnetic media are used to generate copier usage reports that include such information as user identification, number of copies made per machine, and the cost of copying for the designated monitor period. See *copier monitor device.*

electronic keyboard   Refers to a keyboard that is used to generate characters through electronic means rather than through mechanical linkages.

electronic mail   The generation, transmission, and display of business correspondence and documents by electronic means.

electronic stencil cutter   A device employed to produce masters for the stencil duplicating process. The unit reproduces an original document image on special electrical-conducting master material. The image is photoelectronically scanned on a rotating cylinder. Optical images are converted into an electrical signal, which is then amplified and used to create a series of tiny sparks through a stylus. The sparks, in turn, burn holes through the master to form the stencil image. Electronic stencil cutters generate the highest-quality masters for stencil duplicating. See *stencil master; stencil process.*

electronic typewriter   A class of office keyboard equipment that takes its place squarely between office electric typewriters and word processors. Electronic typewriters are relatively inexpensive ($2,500 and under); are compact, utilizing about the same amount of space as an electric typewriter; and offer a full range of features that facilitate text input, though having limited text editing capabilities. As such, these devices are expected to make inroads in traditional office environments that currently use single-element electric typewriters, a market largely dominated by IBM with its Selectric typewriter family.

electropercussive process   A facsimile printing process whereby the facsimile receiver employs a sheet of copy paper covered with a carbon sheet and is mounted on a printing cylinder. As the drum rotates, a single stylus strikes the carbon set with each incoming electrical signal from the transmitter. The force of the stylus causes a carbon mark to appear on the copy paper; by transversing the length of the cylinder, the stylus imprints the entire image of the original.

electrosensitive processor   A facsimile printing process whereby imaging is based on a two-layer paper composed of a white titanium oxide coating and a dark underlayer. The paper, which may be sheet- or roll-fed, is imaged via contact with an electric stylus; as the charged wire touches the paper, the white coating is burned off line by line to correspond to the dark image areas of the original.

electrostatic copier   See *electrostatic process.*

electrostatic plate   An offset plate produced by a copier or platemaker employing an electrostatic process. See *electrostatic process; plate.*

electrostatic printer   A nonimpact printing technique using technology similar to that employed in typical office copiers. It forms a copy by attracting toner particles to a static charge on the surface of a photoconductor, then transferring the toner image to the surface of a sheet of copy paper. In the normal office copier, the charged image (latent image) of the original document is formed on the photoconductor simply through exposure of the photoconductor to reflected light from the document. In an electrostatic printer, the image is formed by a light source (i.e., a laser) which "builds" a static image charge on the photoconductor according to information being supplied through the input data stream. Each bit of data can be related to a character shape in the memory of the printing system, and in most cases characters are formed by a dot-matrix method similar in concept to that of the matrix printer. Paper can be sheet- or roll-fed.

electrostatic process   A copymaking process in which the reflected image of an original document is converted into a static charge that is used to attract toner imaging material to the surface of a copy sheet. In plain-paper copying, the process involves the following fundamental steps: charging of the photoconductor by a corotron; exposure of the original, during which the image area absorbs the light and the nonimage area reflects light; transfer of reflected light via mirrors and a lens to the surface of the photoconductor, leaving a static charge (latent image) which corresponds to the image area of the original; attraction of toner particles to the latent image area on the surface of the photoconductor; transfer of the toner from the photoconductor to the surface of the paper via a transfer corotron; neutralizing of the remaining photoconductor charge by a cleaning corotron and the removal of excess toner. In the coated-paper process the copy paper itself acts as the photoconductor, eliminating the need for a charged intermediate as in the plain-paper process. See *coated-paper process; corotron; photoconductor; plain-paper process; toner.*

element Refers to the interchangeable type font of some impact printing devices; for example, the IBM Selectric "golf ball," Qume or Diablo daisywheel, NEC Spinwriter "thimble."

element printer A class of impact printers which generate copy via interchangeable "elements" that each contain a full set of characters. Characters are formed when the element strikes the paper itself through an ink ribbon. Commonly used element types are "golf ball," daisywheel, and "thimble."

elite type A 12-pitch (12 characters per horizontal inch) typewriter setting. Also, a specific type face.

embossed media Vinylite disk or belt recording media where the recording is made with a stylus embossing needle.

emulation The imitation of one system's code set by another such that the two may communicate. For instance, a system with TTY emulation appears like a teletype system when communicating with another teletype. A system that emulates another system can, in a communications session, accept and process the same data as the imitated system.

endless-loop recorder A dictation system in which a nonremovable magnetic tape is sealed in a "tank" and loops around constantly. See *central dictation system.*

end-of-line decision Deciding whether to put the next word on the same line or carry it to the next line.

end-of-page decision Deciding whether to put a line on the same page or carry it to the next page.

end-of-page stop A feature which stops the printer when it has finished printing a page of text. Usually employed to allow the operator to change paper or printer settings, or to allow the next document to be processed.

engineering size Common document/print sizes in engineering reprographics. They are specified as A-size (8½″ × 11″ or 9″ × 12″); B-size (11″ × 17″ or 12″ × 18″); C-size (17″ × 22″ or 18″ × 24″); D-size (22″ × 34″ or 24″ × 36″); and E-size (34″ × 44″ or 36″ × 48″).

F-size is 34″ or 36″ wide and of indeterminate length.

engineering-size copier   A copier, duplicator, or diazo duplicator capable of reproducing large document originals.

enlargement   The ability of a camera or camera/platemaker to enlarge the image of an original document through the use of a lens with adjustable focal length.

equity   Credit that is accrued from monthly rental or lease payments toward the eventual purchase (if desired) of the leased equipment.

ergonomics   The study of equipment design for the express purpose of reducing operator fatigue and other forms of discomfort, both psychological and physiological, in a man/machine environment.

error-free   A feature of magnetic recording media whereby the user can dictate over any errors in the last phrases.

E-size   See *engineering size.*

etch   (1) An acid solution used to prepare the nonprinting areas of an offset for inking. The solution may be applied automatically by the duplicator, by a manual wiping of the plate surface, or by an etcher. (2) The process by which an offset plate is prepared for inking with an etch solution. See *etcher.*

etcher   A device used to apply an acidified gum solution to the surface of an offset plate to make the nonprinting areas resistant to inking. See *etch.*

European paper sizes   Common paper sizes employed in Europe and Japan. They are designated as sizes A3 (11.7″ × 16.5″); A4 (8.3″ × 11.7″); A5 (5.8″ × 8.3″): B4 (10.1″ × 14.3″); B5 (7.2″ × 10.1″); and B6 (5.1″ × 7.2″).

exception word dictionary   A list of stored words that the word processing system utilizes to perform hyphenation decisions.

expendables   Equipment supplies and machine parts that have a limited lifetime and that must be replaced or replenished on a regular basis. In copiers, such items include toner, developer, paper, webs, brushes, filters, and photoconductors. In duplicators, expendables include ink, fountain solution, etch solution, plates, blankets, and rollers.

exposure   (1) The flashing (illumination) and deflection of a document image onto the photoconductor of a copier or the photographic material of a camera or platemaker. (2) The illumination of a negative or positive brought in contact with a sheet of sensitized paper or film and secured in place by a vacuum frame. (3) The length of time light-sensitive material is illuminated and the amount of light allowed to pass through the copier/camera/platemaker (photo-direct method only) as determined by the aperture setting of the lens or other control (e.g., contrast control on some copiers). The amount of

exposure will affect the resultant lightness/darkness of the reproduced image (shorter exposure creates a darker image; longer exposure creates a lighter image). See *contact print; electrostatic process;
photo-direct exposure.*

exposure contrast control See *contrast control.*

exposure glass See *exposure platen.*

exposure lamp High-intensity lighting employed by copiers, cameras,
and platemakers to expose original documents for reproduction
purposes. In copiers the lamps are either stationary or moving and
are positioned under the exposure platen. In cameras and platemakers the lamps are stationary and generally situated above the
copyboard. See *copyboard; exposure; exposure platen.*

exposure platen The section of a copier on which an original is exposed for reproduction. The platen may be fixed flat-bed or curved
glass plate, or a moving document carriage over a slit exposure system with an integral single-sheet document feeder. See *curved platen;
fixed flat-bed platen; moving flat-bed platen; single-sheet document feeder; slit
exposure system.*

exposure timer (1) A console control dial on a camera or camera/
platemaker which is used to set the duration of the desired plate exposure. (2) A console control dial on a diazo duplicator which is
used to set the paper-feeding speed of the exposure section. The
speed at which the paper is fed will affect the duration of exposure
to the ultraviolet light source.

fabric ribbon Usually a nylon typewriter ribbon.

facsimile A technique used to transmit a literal picture of a document.

facsimile device A machine employed to relay alphanumeric and
graphic data to distant sites along telephone or transmission lines, or
via radio and microwave communication links. A facsimile unit creates a copy in the same manner as an office copier, except that the
"original" to be copied is received electronically, usually over a
phone line. On the transmission end, the original is scanned, converted into electrical signals, then sent to a remote site, where a similar device receives the data and makes a hard copy from it. See also
Group 1, Group 2, Group 3, Group 4.

fan fold Refers to continuous-form paper, used for computer printouts, supplied in convenient flat folded form.

fast forward A tape-recorder feature that permits the tape to be
wound rapidly in normal play direction for search purposes.

feasibility study An investigation of the advantages and disadvantages of using an alternative approach over the presently used approach. In word processing, a study conducted to determine whether
it would be advantageous to convert to word processing.

feed capacity   The size and number of sheets that can be stacked onto the paper feed tray of a copier or duplicator. See *feed tray.*

feed tray   The tray of a copier of duplicator on which the sheet paper supply is loaded for input to the imaging system. See *feed capacity.*

feedback   The ability of a machine to sense and correct its own mistakes. May also be used in connection with systems of communication among members of an organization if the results of communications from the top may be easily assessed by those who issued them.

feeder   The mechanism which indexes the paper supply into the imaging system of a copier or duplicator. See *friction wheel feeder; roll feeder; vacuum feeder.*

fidelity   The degree of accuracy with which an electronic device reproduces sound.

field   A unit of information within a record that serves a similar function in all records of that group (e.g., a personnel record contains a name field, address field, salary field, etc.).

file   In word processing, a segment of text that is callable from storage and is usually one document long. In records processing applications, an organized, named collection of records treated as a unit.

file maintenance   The activity of keeping a file up to date by adding, changing, or deleting information.

file organization   The manner in which files (text segments of paragraph, page, document, or other length) are arranged or formatted and may be accessed on storage media. Also details whether there is an index created automatically for stored text, and whether this index is accessible on the display and via printout.

file select   See *automatic file select.*

file sort   See *automatic file sort.*

film ribbon   Usually a Mylar® carbon typewriter ribbon.

filter bag   A disposable bag found in the imaging systems of some dry toner copiers. It collects excess toner particles that are cleaned from the photoconductor by a cleaning brush.

final copy   A completed (presumably perfectly typed) document.

financial lease   A financing method in which the customer pays monthly rentals to a lessor that in total exceed the purchase price of the equipment involved; the additional cost to the customer represents the lessor's profits and expenses. The customer is responsible for machine maintenance, sales taxes, and insurance. Also called a *full-payout lease.* Financial leases may also be classified as being *true leases* or *conditional sales.*

firmware   A term related to specific software instructions that have been more or less permanently placed into control memory. An ex-

tension to a computer or word processor's basic command (instruction) repertoire to create a user-oriented instruction set (this instruction set is done in read-only memory, not in software). The read-
only memory converts the extended user-specific instruction to the
basic instruction of the system.

first-copy time   The time required for a particular model of copier to
produce a single copy or the first copy in a multiple-copy run.

first-line form advance   A forms-feeding device attached to the word
processor that can be instructed to automatically advance to the top
(first line) of the next form at the completion of a document. This
avoids the need to record keystrokes for multiple-line advances.

fix   See *fuser.*

fixed flat-bed platen   A stationary, planar exposure glass on which
documents are positioned to produce reproductions on a copier. Exposure is from an underlying lighting system, which may be fixed or
moving. See *exposure platen.*

flash cards   Cards on which information is recorded for file reference.

flash exposure   When shooting halftones with a camera or cam-
era/platemaker, a second exposure of the photographic/plate material with a sheet of white paper covering the original on the copyboard. This second exposure, equal in length to 15% or 20% of the
initial exposure, accentuates highlight and shadow dots while keeping middletone dots constant. See *halftone.*

flat-bed platen   See *fixed flat-bed platen; moving flat-bed platen.*

flat comb binder   A type of binding device that combines a two-part
plastic strip (one with spikes, the other with holes) with collated
copy sheets punched along the left margin to form a bound document. The final bind is heat-sealed for permanence. See *binder.*

flexible diskette   See *magnetic media.*

flexible staffing   Use of temporary/casual/part-time employees to
meet peak workloads.

flicking   A term referring to too-rapid use of the Selectric keyboard
whereby random hyphens appear.

flippy   A double-sided diskette.

float   In critical path scheduling, extra time available for a job because
work cannot proceed further until another job that will take longer
is completed.

floppy diskette   See *flexible diskette (floppy)* under *magnetic media.*

flowchart   A graphic representation of the flow of work from origin to
completion in which symbols are used to represent operations and
equipment.

fluid duplicating   See *spirit process.*

fluorescent lamp   See *ultraviolet lamp.*

flush left   A term that describes a block of text that has an evenly justified left margin.

flush right   A term that describes a block of text that has an evenly justified right margin.

FM   Frequency Modulation.

font   A character set in a particular style and size of type, including all alpha characters, numerics, punctuation marks, and special symbols.

font disk   An imprinted glass disk employed to store character fonts in a phototypesetter.

font master   On a photocomposer, the configuration used to store the character fonts, which may include glass font disc, film strip (on a revolving drum), grid matrix, or other less common varieties. A single font master may contain more than one font type.

fonts on-line   The total number of fonts which may be loaded at one time for automatic access on a photocomposer.

footer   Information printed consistently at the bottom of each page of a multipage document.

footing   Adding fields of information vertically.

footnote tie-in   See *automatic footnote tie-in.*

foot pedal   Activates transcribing machine. By pressing the center of the pedal, the operator can listen to recorded dictation. By operating other parts of the foot pedal, the operator can reverse and "fast forward" the rewinding.

forecast   An estimate of performance for a future period (e.g., sales estimate, profit estimate).

foreground processing   A word processing job application such as communications or printing that takes place such that the system remains dedicated to performing that function and cannot be employed to perform another task. Contrast with *background processing.*

format   A contraction meaning the FORM of MATerial, designating the predetermined arrangement of text/data for output.

format decision   Deciding how to set up a page (margins and tabs, line length, etc.).

format statement   Embeds format information with text. The format statement may include such parameters as margin and tab settings, decimal tab settings, centering instructions, paragraph indentations, line spacing, pitch size, etc. Some systems have an automatic or default format, which is used whenever the operator fails to specify a format. Some systems allow only one format per document or per page. Other systems allow the operator to change formats, and automatically call up various formats as desired. Some systems employ a format menu or list instead of a format statement for the document. This is generally stored separately and displayed separately from the document itself.

form letters   The same basic letter to be sent out to a number of different people, usually prepared in advance and duplicated.

form rollers   In the offset process, the rollers in the ink and dampening systems that contact the plate cylinder to distribute ink and fountain solution.

forms feed   Refers to a pinfeed platen or forms tractor device to handle continuous paper for automatic printout.

forms input   Filling in a form, by spacing automatically from field to field with a carrier return or other single action. A few systems define fields as all numeric or all aphabetic, and reject incorrect entries.

FORTRAN   A high-level computer programming language, suitable for scientific applications.

fountain reservoir   The container and tray which supply fountain solution to the dampening system of an offset duplicator. See *dampening system; fountain solution.*

fountain solution   Solution that provides a mildly acidic dampening of the surface of an offset plate to keep the background or nonprint areas of the plate ink-repellant during a printing run. Fountain solution is applied by a roller system from the fountain reservoir to the plate cylinder of an offset machine. Fountain solution may also be referred to as water or dampening solution. See *offset process.*

FPM   Feet Per Minute, as in the exposure speed of a diazo duplicator.

friction wheel feeder   A type of feeder usually employed on tabletop offset, spirit, and stencil duplicators that utilizes rubber-coated rotating wheels to contact the paper on the feed table and propel the top sheet into the impression mechanisms.

full duplex   A communicating word processor's ability to send and receive text simultaneously. Also called *duplex.*

full-equity lease   Purchaser becomes owner of equipment at end of lease.

full-payout lease   See *financial lease.*

function   Usually refers to the uses or ultimate purpose of a given person or thing; for example, a manager performs certain functions, such as planning.

function keys   Keys on a keyboard or a control panel which when depressed activate a particular machine function.

functional classification   Segregation of incomes and expenses by department according to responsibility or control authority.

fuser   In the electrostatic process, the means by which toner is permanently fixed to a sheet of paper by heat and/or pressure rollers. See *electrostatic process.*

FX   Foreign Exchange. A service that connects a customer's telephone to a distant exchange, permitting the customer to make calls to the distant exchange on a local-call basis.

gain control   Automatic adjustment of volume during the recording of dictation so that outside noise is eliminated.

galley proof   Columnar printout of draft copy used for proofreading.

Gantt chart   A chart on which progress in the various parts of a project (e.g., production of a total amount of product, construction of a building) is plotted against time.

GASP   Gas Plasma Display. A video display screen employing a gas plasma or discharge technology, characterized by an exceptionally clear, flicker-free image.

ghost hyphen   See *discretionary hyphen.*

glide time   A timekeeping principle where, within limits, the employee sets his own starting time.

global search and replace   The ability of a system to search for repeated occurrences of a character string (typically up to 32, 64, or 128 characters long). In some instances, the system can automatically delete all occurrences of a string or replace all occurrences of one character string with another character string. In other cases, the system merely locates the string for operator-selected deletion or replacement. A few high-powered systems can apply logical considerations to making the replacement or perform multiple searches simultaneously.

glossary   See *term dictionary/glossary.*

goal   That object or condition with respect to which the behavior of individuals or groups is desired. The terms "reward" and "incentive" are sometimes considered as equivalent. However, in connection with organizations, the term "goal" is usually applied to a prescribed objective of work, whereas rewards and incentives are need-fulfilling things or conditions, received or attained by the person through working. Ideally, the attainment of goals should result also in reward for the individual.

government size   Paper-sheet sizes commonly used by the U.S. government. They are $8'' \times 10\frac{1}{2}''$ (letter) and $8'' \times 13''$ (legal).

grade   The quality of paper as determined by the components of the stock (wood fiber, cotton fiber, etc.) and the consistency of processing.

gripper margin   In offset duplicators that use chain delivery, the leading edge of the printed sheet contacted by the grippers. This marginal area, which is usually between ⅛" and ¼" wide, will not be printed by the duplicator. See *chain delivery; grippers.*

grippers   Metal bars employed on chain delivery systems in offset duplicators to grasp each individual printed sheet and guide it to the delivery tray. See *chain delivery.*

Group 1   CCITT classification of analog facsimile devices which operate at the speed of 6 minutes using FM modulation. A 4-minute speed is also included in this category as a manufacturer's option.

Group 2   CCITT classification of analog facsimile devices which operate at a speed of 3 minutes using AM modulation. A 2-minute speed is also included in this class as a manufacturer's option.

Group 3   CCITT classification of digital facsimile devices which operate at 1-minute speeds and employ run-length coding of image material to perform redundancy reduction. These machines may also utilize bandwidth compression to enhance speed.

Group 4   CCITT classification of special types of high-speed (56Kb per second) machines with wide scanners used for the transmission of large documents.

grouping   The combining of multiple secretarial areas to facilitate secretarial support for principals.

half duplex   The sending or receiving of text by a communicating word processor in one direction at a time.

halftone   A reproduction of a continuous-tone photograph by a camera or camera/platemaker equipped with a fine-grid glass screen mounted in the lens assembly. The screen acts to break up the artwork into a minute dot pattern, which is transferred to an offset plate. The dot pattern is more suited to offset reproduction than an unscreened photograph.

halftone screen   A film or glass contact screen employed for producing halftone reproductions. The screen contains opaque cross-ruled lines through which reflected light from the original must pass before exposing the film material. The resolution of a screen is determined by the number of parallel lines per inch, and may range from 65 to 150 lines/inch for photograph reproduction. Also known as a *levy screen.* See *halftone.*

handshake   In communications, a preliminary exchange of predetermined signals performed by modems and/or terminals to verify that communication has been established and can proceed.

hard copy   Machine output in a permanent, visually readable form for human beings; for example, printed reports, listings, documents, and summaries. The term has gained significance in light of the use of magnetic records which cannot be read by humans and require processing for conversion to printed records, or CRT display which is transient.

hard-sectored A term used to describe a particular diskette format and a way of recording information on the diskette. Hard-sectored diskettes employ a single index hone placed between any two of 32 equidistant sector holes. The index hole is used to designate the beginning of the disk; the sector holes designate the location of information on each disk. Since hard-sectored diskettes are not preformatted, they have more potential storage capacity than the soft-sectored variety, employing up to 300K bytes out of a possible 400K for text storage.

hardware The mechanical or electronic equipment which is combined with software (programs, instructions, etc.) to create a word processing system.

hardwired A WP system employing wired circuitry to implement system functions. Such equipment is generally cheaper than software programmed systems; it is also less flexible.

head See *read/write head.*

header Information printed consistently at the top of each page of a multipage document.

headers/footers See *automatic headers/footers.*

headliner A photo lettering machine with the capacity to produce headline styles.

heavy-duty I/O (input/output writer) A large Selectric typewriter with special parts added to hold up under heavy usage.

high-level language A programming language that allows the programmer to express operations in a form that is closer to the normal human-language representation of the procedures the computer is to perform. Common examples are COBOL (for business applications), FORTRAN (for mathematical work), and BASIC (an easy-to-use language). See *compiler, machine language.*

highlighting The ability of a display-based word processor to intensify, blink, or create a reverse video image on the display to emphasize a text segment designated for some system activity such as delete or move.

historical data Information concerning past production in a department or any data that have been accumulated for prior periods.

Hollerith code An alphanumeric punched-card code invented in 1899.

horizontal files File storage organized sideways rather than front to rear.

horizontal scrolling Ability of a display-based system to move horizontally along a line of text to access more characters than may be shown on the screen at one time. Several methods may be used. The system can move horizontally across the line, adding one character at a time, or it may display the text as overlapping left, center, and right segments, etc. Some systems display wide lines by condensing character size so that a large number of characters may be displayed.

hot melt/glue binder   (1) A type of binding device which applies melted glue pellets to the spine of a multipage copy set to form a bound document. (2) A type of binding device which applies a solid strip of glue to the spine of a multipage copy set. The copy set and glue strip are placed inside a wraparound cover, and heat is applied to melt the glue and seal the bind. See *binder.*

hot zone   A sometimes adjustable area at the right-hand margin, controlling placement of the last word on a line of text. Most systems wrap any word which will not fit within the hot zone to the next line; others stop for a manual hyphenation decision. See *hyphenation.*

hunting service   Hunting allows the records to be set up so that if one is in use, the originator is connected to the next one in a number series.

hyphenation   Techniques employed on or by a word processing or photocomposition system to perform line ending decisions. Common hyphenation procedures include manual hyphenation, where the operator assigns the hyphen position; hot-zone hyphenation, where any word that enters but does not fit within a predefined end-of-line space must be either manually hyphenated or moved to the next line; and scan, where the system indexes through the text and stops at any place where a hyphenation decision is required. Some systems perform no hyphenations (hyphenless justification) but wrap any word which will not fit entirely on one line to the next line; generally, the operator can override this wrapping and perform a manual hyphenation to maintain tight lines. Discretionary hyphenation requires that the operator key in hyphens at appropriate break points in long or complicated words. The hyphen is recorded by the operator, usually during the initial keyboarding, and is used during subsequent processing only if the word appears at the end of a line; the system will then use one of the discretionary hyphens at the end of the line, justify the line, and move the remainder of the word to the beginning of the next line. More sophisticated systems may use an algorithm (formula) to make hyphenation decisions, or may store a dictionary of hyphens to hyphenate automatically.

hyphen drop   When using an editing typewriter, if a hyphen has been recorded to divide a word at the end of a line, that hyphen should not print if the word ever appears in the middle of a line in adjusted playback. Most units perform this task automatically but will retain required hyphens when instructed.

hyphenless justification   See *hyphenation.*

imaging, image   Creating an impression on paper.

imaging system   The mechanism employed by a copier to create reproductions from an original document. See *dual-spectrum process; electrostatic process; thermographic process.*

impact printer   Any type of printer that generates characters by using some form of stamping or inking through a ribbon by some sort of character slug, element, or hammer-needle. See *band printer; chain printer; cylinder printer; drum printer; element printer; matrix printer.*

implementation   The phase following the approval of the word processing system, during which the details of the new system are developed and carried out.

implementation schedule   A listing in chronological order of the steps involved in conducting a word processing study.

impression   A single printed image formed by a duplicator as the blanket cylinder comes in contact with the copy sheet. See *offset process.*

impression cylinder   The cylinder on an offset duplicator that "backs up" and propels the sheet of paper as it passes between it and the blanket cylinder to receive the inked image from the blanket cylinder.

impression system   The mechanism employed by a duplicator to create reproductions from a master/plate. See *direct lithographic process; offset process; spirit process; stencil process.*

index   A list of documents contained on a unit of storage media (e.g., a diskette). Also, a list of documents being manipulated (such as in a printout queue).

index of cooperation   In facsimile, the length of an individual scan line. To reproduce a document precisely, the index of cooperation should be the same for the two communicating machines.

index slip   A strip of paper containing information about the contents of the recording (number of items, length of each item, and special corrections).

indexing   To move down vertically.

individual loop   One trunk housing a continuous-loop magnetic tape to receive dictation from one dictate station.

information   The meaning derived from the relationship among symbols (words, data).

information processing   A term that encompasses both word processing and data processing, and is used to describe the entire scope of operations performed by a computer.

information system A group of computer-based systems and data required to support the information needs of one or more business processes.

in-house printing See *in-plant printing.*

in-house reprographics The printing or reproduction of documents within a company or organization rather than by an independent printing concern.

ink form rollers See *form rollers.*

ink fountain The tray that holds the ink supply in an offset duplicator. Ink is transferred from the tray to the plate cylinder via a series of rollers. See *form rollers; offset process.*

ink-jet printer A nonimpact printing technique which utilizes droplets of ink to form copy images. As the print head moves across the surface of the copy paper it shoots a stream of tiny electrostatically-charged ink drops at the page, placing them precisely to form individual print characters.

in-plant printing Production of documents by printing or duplicating methods within the user organization, rather than by a separate printing concern. Often used to refer to an in-house facility that has photocomposition capabilities.

input The data or text to be processed. Also, the transfer of data or text to be processed via a keyboard or external device to an internal storage device.

input device A device such as a CRT/keyboard, OCR scanner, etc., which converts data from the form in which they have been received into electronic signals that can be interpreted by a word processor.

input media The various forms and methods of inputting material to a photocomposer for typesetting, including *direct input, paper tape, magnetic media,* and some special applications such as *unjustified paper tape* and *wire service tape.*

input processing equipment Dictation equipment used to supply material to the word processing center.

input underline See *automatic input underline.*

integral keyboard A wood processor or photocomposer with a built-in keyboard for text and command input purposes.

integration A term used to describe the phenomenon whereby different computer-based functions such as word processing, data processing, and telecommunications are capable of being performed by one system.

intelligent terminal A terminal with some logical capability; a remote device which is capable of performing functions upon input and output data.

intensifier See *toner intensifier.*

interaction Reciprocal reactions of people in a group to each other, or of variables in a situation or machine.

interactive Pertaining to an application in which each entry elicits a response. An interactive system may also be conversational, implying continuous dialog between the user and the system.

interactive operation On-line operation where there is a give and take between person and machine. Also called "conventional" mode.

intercharacter spacing Placing white space between the characters of individual words in order to create a justified (even left- and right-hand margins) column of text. Some systems have very sophisticated schemes, with spacing assigned according to character width (even with nonproportionally-spaced fonts), creating a printlike appearance.

interface A shared boundary defined by common physical interconnection characteristics, signal characteristics, and meanings of interchanged signals.

interleaver A device used to insert blank pages between printed pages in a duplicating delivery cycle to avoid the "set-off" of ink from the front of one printed page to the back of another.

interpreter Software that translates program instructions written in a high-level language such as BASIC, FORTRAN, COBOL, or PL/I and possesses the ability to execute the instructions, one step at a time. The presence of an interpreter permits a user to execute previously written programs or to develop his own programming.

interword spacing Placing white space between words to create justified, even left- and right-hand margins in columns of text.

investment tax credit A credit against income taxes, offered by the IRS as an incentive for companies to acquire new equipment. The one-time credit is equal to 7% of the investment if the depreciation period for the unit is seven or more years; two-thirds of the 7% if the lifetime is between five and seven years; and one-third of the 7% if the lifetime is between three and four years.

I/O Input/Output device.

IPH Impressions Per Hour, as in duplicating.

IPN Information Processing Network.

IPS Inches Per Second. Usually refers to speed of tape moving past a recorder head. Faster speeds generally produce higher fidelity.

IWP International Information/Word Processing Association. The largest professional organization for word processing managers and other word processing professionals.

jogger A device that aligns sheets into an evenly piled stack. It is operated off-line from a copier or duplicator.

justification In word processing, the ability of the system to produce printout with an even right-hand margin. This may be achieved by interword spacing (leaving extra white space between words), or by intercharacter spacing with proportionally spaced characters, which provides output with a more printlike appearance. In photocomposition, the adjustment of a line through word and character spacing so that it will exactly fill a given line measure and will have margins (left and right) that are the same as those for other lines in the same measure. See *line measure.*

justify   To output text with flush left and right margins (and, hence, a more printlike appearance).

K   Kilo. Abbreviation denoting 1,000 units. Generally used with a numeric prefix. In "computerese," denotes 1,024; thus 32K bytes of memory would equal 32,768 bytes.

kerning   In photocomposition, the ability to add or subtract space between characters, and expand or compact the word and line. Also known as mortising, white-space reduction, or white-space expansion.

keybar   The conventional kind of typewriter typebar mechanism.

keyboard copying   Copying a page by retyping it.

keyboarding   Entering information to a word processor via a keyboard.

key counter   See *card or key station counter.*

key operator   A person among the users of a copier who is assigned to maintain the machine. Key-operator responsibilities usually include the replenishment of toner, paper, and other supplies as well as the correction of paper jams and minor malfunctions.

key-to-disk   By typing on the keyboard, the operator records the information directly onto a disk.

key-to-tape   By typing on a keyboard, the operator records information directly onto a magnetic or paper tape.

KSR   Keyboard Send/Receive.

KTS   Key Telephone Service. Operates through pushbuttons that connect phones to several lines.

lag   A delay in reactions; specifically, the delay in adjusting costs to a change of conditions.

language skills   Correct spelling, punctuation, grammar, and diction.

large-document copier   A copier, duplicator, or diazo duplicator capable of reproducing large document originals.

laser   Light Amplification by Stimulated Emission of Radiation. A device which transmits an extremely narrow and coherent beam of electromagnetic energy in the visible light spectrum. Lasers have numerous applications in many fields; in the office automation area they are already applied in communications, facsimile, storage, and electrophotographic printing.

latent image   In electrostatic printing/copying, a static charge present on the photoconductor prior to contact with toner particles. See *electrostatic process.*

leading   In photocomposition, spacing between lines and paragraphs, expressed in point and half-point values (pronounced *ledding*). This is one of the primary typesetting control parameters, and refers to the amount of film or photopaper advancement between lines of composed text. See *point.*

leading end clamp   The clamp on a plate cylinder which secures the top edge of an offset plate to the surface of the cylinder.

lease-purchase plan   A lease plan in which monthly payments are applied toward the purchase of equipment. See *conditional sale.*

leasing   Renting plant or equipment as opposed to purchasing. A wide variety of leasing arrangements is available.

LED   Light-Emitting Diode. A form of display lighting employed on many different types of office and reprographics equipment.

ledger size   A sheet size commonly used by business for accounting purposes: $10'' \times 14''$ or $11'' \times 17''$.

legal size   A sheet size corresponding to the standard size of legal briefs, which is $8\frac{1}{2}'' \times 14''$. Legal size has been adopted as a common large size for general business purposes.

lessee   The customer or user of a lessor's equipment.

lessor   The owner of the equipment that is being leased; usually a third party who purchases the item from the original manufacturer, and then offers it to a user for a monthly rate.

letter-quality printer   A printer that generates output that is suitable for high-quality business correspondence. Term implies that output quality matches that of a standard office electric typewriter.

letter size   The common letter sheet size used by business: $8\frac{1}{2}'' \times 11''$.

letter spacing   See *kerning.*

letter writing   See *automatic letter writing.*

level   Used in paper-tape or magnetic-tape jargon, refers to vertical rows of perforations or electronic codes.

levy screen   See *halftone screen.*

licensing   Granting a license to manufacture a trademarked and/or patented product to another company in return for a royalty on each unit sold.

line   Transcribing work is most commonly measured by number of "lines," a standard line being a six-inch line of elite type (12 characters to the inch) or 72 typewritten strokes. Because "lines" could include rough drafts, etc., such statistics would be meaningless. Therefore, only "net" lines (those lines of finished typing ready for dispatch) are usually counted. Allowances are made for headings, endings, carbon copies, and envelope addressing.

linear programming   A mathematical technique for determining the optimum in cases where the relationships between the variables are linear.

line justification display   The ability of a word processor display to show justification (even right-hand margin). Most display screens show this via interword spacing (extra "white" space between words). A few displays have a proportionally spaced character set, and can display justified lines of proportionally spaced characters with justification occurring via intercharacter spacing.

line measure   The width of the line of text on a composed page, expressed in picas. See *pica.*

line printer   A computer or word processing output peripheral which operates at a very high speed to print what appears to be one full line at a time; with matrix printers that is accomplished through several sweeps of the print head.

line spacing   See *automatic line spacing; leading; reverse leading.*

line speed   The rate at which text is transmitted over a line, expressed in bits per second.

liquidated damages charge   A charge levied by a supplier to a customer when the customer cancels a rental contract prior to the agreed term of the plan.

liquid toner   See *toner.*

liquid-toner process   An electrostatic copying method employing toner material composed of carbon particles suspended in a liquid carrier. Distribution of the liquid toner to facilitate image production may be via a photoconductor drum intermediate (plain-paper process) or directly to the copy sheet (coated-paper process). See *electrostatic process; toner.*

list processing   See *records processing.*

lithography   Also called photo-offset lithography, offset lithography, or offset. See *offset process.*

load   In word processing, to feed a program into the system. A common means of loading the program is via a form of magnetic media. The medium is inserted into the media drive and the program is "read" into the system's memory. Also referred to as "soft loading." See *software-based.*

load transfer   A method of rerecording material from a continuous-loop recorder onto a different medium.

lockout   A mechanical or electronic method of ensuring that dictation system users will not be able to intrude on anyone else's recording.

logging   A method of recording, cataloging, and/or filing documents or media.

logic   Instructions programmed into a unit to control how an operation is to be done.

log sheet   A document prepared and maintained by operators to index work on any given tape, card, etc.

lower case   Small type letters in contrast to capital letters or upper case.

LPH   Lines (of typing) Per Hour. Unless specified, this should always refer to "net" lines of finished typing ready for final dispatch or disposition.

LPM   Lines Per Minute, relative to composition speed on a phototypesetter.

LQP   See *letter-quality printer.*

LSI   Large-Scale Integration. Refers to a microprocessor chip with more than 1,000 components.

M   Mega. Abbreviation denoting 1 million units; 10M bytes denotes 10 million bytes of storage in "computerese."

machine dialog/menu prompt   A number of video display systems allow the operator to enter into a dialog or conversation with the word processor; operator actions are called for by the system in a question–answer mode. This type of dialog can be particularly useful in training new operators, or in helping part-time operators through a job. Some video systems offer menus which list operator options at each step in the word processing process.

machine language   A binary language all digital computer products must use to perform processing. All other programming languages (BASIC, FORTRAN, COBOL, etc.) must be compiled or translated ultimately into binary code before entering the system. This binary language is used directly by the machine.

machine shorthand   See *stenotype.*

machine transcription   To make a typewritten copy of dictated material using a transcribing machine. The distinctive feature of machine transcription is that the typist works not from visual materials, such as shorthand notes, but directly from a voice recording.

mag card   See *magnetic media.*

mag card/mag tape   Tape or card coated or impregnated with magnetic material, on which information may be stored in the form of tiny magnetic spots.

mag keyboard   See *magnetic keyboard.*

magnetic brush   In the electrostatic process, a device used to transport toner particles to the surface of the photoconductor. A brush is mounted on a roller mechanism. As it turns, its magnetic pull attracts toner particles and carries them to the drum surface, where they are deposited. See *electrostatic process.*

magnetic cartridge   See *magnetic media.*

magnetic keyboard   A word processor. The term implies that the system is provided with magnetic media to capture keystrokes, to record text, and to permit revisions.

magnetic keyboard output   Playback of previously recorded typing from magnetic tape or card initiated by an operator or prerecorded command.

magnetic media   A variety of magnetically coated materials used by word processing systems for text and sometimes program storage. Main types of magnetic media include:

• **magnetic card**   Tab-size card coated with magnetic material, holding about 50 to 100 lines (about 100 characters each) of text and codes. Dual-sided cards are also available (recording on both sides). Another, less frequently used, card type is the mag stripe card, which holds about a paragraph of text.

• **cassette**   Magnetic tape loaded into a reel-to-reel cassette with a capacity of approximately 30 text pages.

• **cartridge**   Magnetic tape loaded into a cartridge (such as the single-reel IBM MT/ST cartridge or the reel-to-reel 3M Data Cartridge) that holds multiple pages of text.

• **flexible diskette (floppy)**   Magnetic coated Mylar® disk enclosed in a protective envelope. Three basic sizes:
—Standard diskette: 8″ diameter, capacity of approximately 75 text pages.*
—Minidiskette: 5¼″ diameter, capacity of approximately 15 to 20 text pages.*
—Microdiskette: Developed by Sony, enclosed in a plastic case. 3½″ diameter, capacity to store 278K characters when formatted (437.5K unformatted).

• **disk**   Rigid, random-access, high-capacity magnetic storage medium. Disks may be removable (cartridges), providing off-line archival storage, or nonremovable. Capacities range from 1Mb to well over 300Mb (250 to 750,000 pages*) per disk.

• **Winchester disk**   Rigid, nonremovable, magnetic oxide-coated, random-access disk sealed in a filtered enclosure along with the read/write heads and head actuator. Heads fly only about 20 micro-inches from the disk surface, allowing very dense data storage. Common disk sizes are 8″ and 14″ in diameter. Capacity ranges from 2.1Mb to 64Mb (325 to 16,000 pages*) for the 8″ disk, and from 6.5Mb to 635Mb (1,625 to 158,750 pages*) for the 14″ disk.

* Calculated using 4,000 characters per page, single-spaced. Storage capacity can be increased through use of a double-density diskette, a double-sided diskette, or a double-density, double-sided diskette.

magnetic-media typewriter   A typewriter on which information is recorded and stored on a magnetic medium for distribution to a distant location.

Mailgram   A letter sent by Western Union and delivered within 24 hours.

makeready   Making necessary preparations to begin a task (loading paper, changing font, etc.).

make ready/do/put away   The basic components that make up any job.

manual selector   A manual switch on the dictate station enabling the dictator to select a certain dictation recorder.

margin adjust   See *automatic margin adjust.*

marginal analysis   Analysis to determine the point at which the cost of extra input of one factor (e.g., another employee or another machine) will pay for itself.

marginal income   That portion of income which remains after variable expenses are met, and is therefore available for fixed costs and profit. (Variable expenses are those which are proportionate to and incurred because of the receipt of the related income.) Marginal income may be expressed variously; for example, as a percentage of the sales dollar or as the number of dollars remaining after the variable expenses of a given volume of sales are met.

master   (1) An offset plate. The term usually refers to a low-volume plate not good for more than 1,000 impressions. (2) A plate used to generate impressions on a spirit or stencil duplicator. See *plate; spirit master; stencil master.*

master budget   A summary report including the program of operations, budget formulas, profit–volume relationships, the capital-expenditures budget, the cash budget, and the budgeted balance sheet of an entire company or major division.

mastermaker   See *platemaker.*

matrix printer   An impact printer which uses wire, hammerlike bristles or needles to create characters formed by small dots. Matrix printers produce either serial or line output. The serial printer employs a moving print head with a matrix block (5×7 or 7×9) of needles. The print head sweeps across the page to print full characters one at a time. The line printer uses a horizontal band with raised dots which moves from left to right across the paper. The individual needles strike programmed character dots to form one row of dots per sweep across the page. Successive passes of the line printer form complete characters and complete rows of textual data. High-resolution text, comparable to daisywheel output, may be produced by overlapping matrix printers which print characters via a highly concentrated matrix or successive, staggered passes of the print head. Fonts for matrix printer are stored in ROM or PROM memory.

maximum image area The maximum copy image or print area capable of being produced on a particular copier or duplicator, regardless of the maximum sheet size possible.

maximum line length In typesetting, the maximum width of a line of type which may be set by a typesetter model. This is expressed in picas and is typically about 45 picas.

maximum sheet size The maximum sheet size capable of being used on a particular copier or duplicator.

MC/ST IBM Mag Card/Selectric Typewriter.

measured backspace A transcriber device where, according to user preference, there is a measured repeat of dictation recorded previously each time the foot pedal is depressed.

measured review See *measured backspace*.

media General definition of the various types and kinds of things on which recordings can be made.

media compatibility The ability to share media among secretaries who are receiving the media from different systems and even from different vendors.

media converter A device, or the capability of a word processor, to convert information stored on one type of media to another.

media-stored format A feature of a word processing system that allows format information (such as tab and margin settings, paragraph indentations, etc.) to be stored on magnetic media with the text, and to be used to format text on the display and/or during printout.

memory In word processing, usually refers to a self-contained RAM memory like core or MOS (metal oxide semiconductor).

memory tape In copying, a special form of photoreceptor belt which retains a latent conductivity pattern for repeat image transfers from a single exposure of an original. The conductivity image (charge) does not immediately dissipate like an electrostatic pattern, it has a natural decay cycle, which allows up to ten toner-image transfers to plain paper. The tape may be reused from three to five times after regeneration in total darkness for 48 to 72 hours. The photoreceptor is composed of a plastic belt base coated with titanium dioxide (TiO^2). 3M currently employs this technology on the VHS-R plain-paper copier.

memory typewriter A typewriter that is capable of storing keyboarded material and playing it back automatically. Memory typewriters generally have some text input features and compete in the low end of the word processing market.

menu A list of alternative operator actions, supplied by the word processing system for operator selection. In some cases, the system will require that the operator access some or all functions through the appropriate menu.

mercury vapor lamp An ultraviolet light source commonly used in the exposure systems of diazo duplicators. See *diazo process.*

merge See *document assembly/merge.*

messengers Personnel trained to deliver mail and documents on a regular schedule between principals and the word processing center.

metal plate See *plate.*

microcassette Miniature cassette smaller than a minicassette.

microfiche A sheet of film or a card containing multiple micro-images in a grid pattern. It usually contains identification information which can be read without magnification. Also called "fiche."

microfilm A film on which printed pages are photographed in reduced size for convenient storage and use.

micrographics Combines the science, the art, and the technology by which information can be quickly reduced to the medium of microfilm, stored conveniently, and then easily retrieved for reference and use.

microphotography The process of making miniature copies of printed or other graphic material.

microprocessor An integrated circuit which contains the logic elements for manipulating text/data and performing processing operations on it.

microwave High radio frequencies, nominally between 1,000 and 200,000 megahertz.

mil One one-thousandth of an inch. Generally refers to tape thickness.

mimeograph See *stencil process.*

minicassette Miniature cassette. Smaller than standard cassette but larger than microcassette.

minicomputer A small computer with the same kinds of capabilities as a large computer, but on a limited scale.

mini floppy diskette See *magnetic media.*

misfeed The failure of paper to be delivered from a copier or duplicator due to an obstruction in the paper path, machine malfunction, misloading of the paper feed supply, or copy paper jamming because of excessive paper curl or sheet damage.

mix The specific quantities of the various components comprised in a whole; for example, the sales mix expresses the quantities of each product included in total sales.

mnemonic The assisting of the human memory. In word processing this refers to a type of command structure the system may employ. A system that utilizes mnemonic commands uses commands that are often abbreviations for the function they implement—for instance, a "CO" command used for a text copy application.

mode The operating conditions of a unit.

model, mathematical A description of a system (a machine or a situation) in which the factors are expressed in mathematical terms in such a way that effects of various changes in the real system can be simulated by changing the values given to the mathematical terms.

modem Contraction of MOdulator-DEModulator. A device which modulates and demodulates signals transmitted over communications facilities; that is, a device used to convert digital signals into analog (voicelike) signals for transmission over a telephone line. At the other end of the line, another modem converts the analog signals back into digital form. A modem is also known as a data set.

modified Huffman code The run-length code adopted by the CCITT to perform redundancy reduction; it removes only horizontal redundancy from the image. CCITT is currently studying a two-dimensional run-length code which removes both horizontal and vertical redundancy. See *run-length code.*

module An interchangeable plug-in item or a compatible add-on unit.

moisture development A type of developing process employed on diazo duplicators. The coating of the copy material is composed of diazo compound and a stabilizing acid barrier; couplers are contained in a liquid alkaline development solution. Following exposure, the copy paper is passed between developer applicator rollers that spread a metered amount of solution over the sheet to bring out the copy image. The system is odorless and requires no venting. See *diazo process.*

monitor device See *copier monitor device.*

Monte Carlo techniques Operations research techniques that make use of the laws of probability.

motion study Study of the discrete motions performed in doing a task to determine whether the task can be performed more efficiently.

mottle On copied or printed sheets, the spotty or uneven density of line or solid areas.

moving flat-bed platen A planar exposure glass mounted on a moving carriage on which documents are positioned to produce reproductions on a copier. Exposure is from an underlying slit exposure system, over which the platen moves. See *exposure platen.*

MSR Marketing Support Representative. IBM term for a person who calls on customers to assist the sales representative in selling word processing equipment, performing vendor feasibility studies, training customer operators, and bringing new applications on-line.

MTM Methods Time Measurement. A measurement that recognizes certain basic motions, which occur repeatedly in varying combinations and sequences in the performance of all types of office and factory work.

MT/SC IBM Magnetic Tape/Selectric Composer.

MT/ST IBM Magnetic Tape/Selectric Typewriter.

multicolumn page display The ability of a word processor to display text or numerics in multiple columns. The system may also have the ability to edit by column. See *column move/delete.*

multidrop line A communication system configuration using a single channel or line to serve multiple terminals. Use of this type of line normally requires some kind of polling mechanism, addressing each terminal with a unique identification. Also called a multipoint line.

multifunctional Able to perform more than one function.

multileaving A technique for allowing simultaneous use of a communications line by two or more terminals.

multimedia system A system that uses more than one kind of medium, such as both cassette and endless loop for dictation recording.

multiple-copy rate The speed at which a copier will operate in an uninterrupted continuous copying mode.

multiple-part form A form with duplicate pages underneath the top sheet so that information typed on the top sheet is duplicated on the copies through carbons.

multiplexer A hardware device that allows the transmission of a number of different signals simultaneously over a single channel.

multiplexing The division of a transmission facility into two or more channels either by horizontally splitting the frequency band transmitted by the channel into narrower bands, each of which is used to constitute a distinct channel, or by allotting this common channel to several different vertical information channels one at a time.

multipoint line See *multidrop line.*

multiprogramming A word processing system that can simultaneously process different applications such as file sorting and text editing or even data processing tasks. Systems with such capabilities can process much larger quantities of material and are more cost-efficient.

multiterminal system See *shared-logic word processing system; shared system.*

natural language In word processing, refers to a system whereby functions are implemented through a keyed sequence that may combine verb, noun, and object commands, along with numerics to imitate English language syntax. For example, to delete the ninth paragraph within a document the operator would depress the "Delete + Paragraph + 9" keys. Such systems may theoretically facilitate the training of operators, since the commands employed closely resemble the English language.

negative corotron   See *corotron.*

net lines   Refers to "finished" lines of typing ready for dispatch or final disposition. Includes author corrections.

network   See *communications network.*

nonadjust printout   The ability of a word processing system to print material with protected (unchanged) line lengths; this is required to retain the format of tables, charts, and other tabular material.

noncounting   A composition system which does not contain internal logic to perform line justification. The term generally refers to an off-line editing system that is used to compose unjustified text for input to a typesetter. See *justification.*

nonimpact printer   A class of printers that form images onto paper without using a form of stamping or inking through a ribbon by an element, character slug, or hammer-needle. The shapes of characters are stored in the system memory of the printer and used to drive the imaging mechanism. See *electrostatic printer; electrostatic process; ink-jet printer; thermal printer.*

nonlinear programming   An operations research technique that may be used when some of the variables in the mathematical models are raised to a higher power (e.g., x^2).

nonselector   One recorder only for a dictation system; recorder cannot be selected by the dicatator.

nonswitched line   A communications link which is permanently installed between two points.

nonvolatile   A type of memory which holds text/data even if power has been disconnected.

numberer   An offset-press attachment which automatically imprints consecutive numbers on printed material (i.e. checks) during the delivery cycle of the paper path.

numeric character   A character that belongs to one of the set of digits 0 through 9.

numerical control   Direction of machines by magnetic tape or punched cards.

NWAM   Net Words A Minute.

objective classification   Segregation of expenses according to the nature of the expenses; that is, by type of expense such as labor, supplies, etc.

obsolescence clause   Allows purchaser of WP equipment to trade in equipment when newer model comes out. Manufacturer agrees to buy back present equipment at an agreed price. The full buy-back amount is credited against the purchase price of the new equipment.

OCR   Optical Character Recognition. A device or scanner which can read printed or typed characters and convert them into a digital signal for input into a data or word processor. OCR units in word processing applications usually read special machine-readable type fonts (OCR-A, OCR-B, or the IBM Courier font). The use of such equipment allows an ordinary typewriter fitted with a font to serve as an input station for a word processing system. Pages produced on the typewriters are fed into the OCR and converted into a digital form. Such digitized text may either be entered directly for text edit and format, or stored on magnetic media for future processing.

OEM   Original Equipment Manufacturer. An OEM may manufacture a product for assembly into another system or larger configuration by another manufacturer or vendor.

off-line   A word or data processing operation performed on stand-alone equipment, not being connected to another central processor or computer system.

off-line keyboard   In typesetting, a keyboard input device that is not directly wired or connected to the typesetting component of a composition system. Text and commands can be entered and recorded off-line for subsequent entry to a typesetting device via a transferable medium like paper tape or floppy disk. An off-line keyboard may or may not be capable of performing line justification with text input.

offset   (1) In duplicating, when the bottom side of a delivered sheet picks up the inked image from the sheet below it in the stacker. Also called set-off. (2) A sheet stacking configuration in which the delivery mechanism alternates the stacking direction of printed sheets. This technique may also be applied to the delivery system of a sorter, allowing multiple-page sets to be stacked in an alternating fashion. (3) The offset printing process. See *offset process.*

offset duplicator   See *offset process.*

offset lithography   See *offset process.*

offset process   A duplicating technique based on the fact that oil (grease) and water do not mix, and that impressions may be "offset" from a plate to an intermediate and then finally onto a page. Offset plates have ink-receptive image areas and water-receptive clear backgrounds. The plate or master is first "inked" on a plate cylinder by a number of ink and water solution rollers. The image is then transferred from the plate onto an intermediate rubber blanket cylinder. Finally, the image is transferred or offset from the blanket onto a blank page that is backed up by an impression cylinder. Inks employed on offset duplicators are oil- or grease-based, and are specially formulated to adhere to plate images. See *plate.*

one-for-one structure   An office structure in which the secretary performs both administrative tasks and production tasks for a single boss.

onionskin paper   Lightweight stationery used to make carbon copies.

on-line   A word or data processing operation which is performed on a local system connected to and sharing the facilities of a remote central processor.

on-line sorter   A sheet sorter that is configured on-line with a copier or duplicator. Copies are carried from the delivery system of the reproduction unit directly into the feeder of the sorting device. See *sorter*.

open landscape   Refers to office design in which there are no walls and partitions separating work areas.

operating lease   A short-term lease in which the lessor retains ownership of the equipment. Rates are usually lower than the monthly rentals offered by the manufacturer. Under an operating lease, the lessor does not expect to recover the full cost of the equipment from one customer. Therefore, when a lease expires, the lessor must rely on continuing the relationship with the same user, or find another customer. The rates may reflect the risk involved. Also called a partial-payout lease.

operating system   Software that controls the operation of the word processing system.

operations research   Application of quantitative methods to the solution of management problems. It includes a "systems approach" in which a large number of factors, rather than a few, are considered positively by a group of scientists from different fields.

organizing   The managerial function which involves identification of tasks to be accomplished; description of duties of various positions; recruitment and assignment of proper individuals to these positions; and, in general, design of all the administrative instruments needed to accomplish the goals of the unit.

original   A document to be reproduced by a reprographics system such as a copier or duplicator.

originator   A person who creates ideas to be typewritten onto a page.

oscillating scan head   A scanning process that employs an optics/photocell assembly that moves back and forth across the original page as it scans each line. The document is mounted on a semicylinder platen which, with each oscillation, moves laterally beneath the scan head.

output   The product of an information processing operation, produced via display or a peripheral device such as a printer, communications, mag card reader, etc.

overhead   Costs that do not vary with output over a period of time.

PABX   Private Automatic Branch Exchange. A private automatic exchange that provides for the transmission of calls to and from the public telephone network.

packet   A group of bits including data and control elements which is switched and transmitted as a unit. The data are arranged in a specified format.

packet switching A mode of transmission in which a message is divided into fixed-length packets within the network, as opposed to message switching systems, which route a message in its entirety. Packets are routed over the network under computer control to take advantage of the speed with which shorter messages may be transmitted and reassembled at the receiving end.

page numbering See *automatic page numbering.*

pages See *messengers.*

page scrolling The ability of the system to "flip" through the pages of a document, usually in both forward and backward directions, allowing access to all text of a multipage document.

pagination See *automatic pagination.*

paper misfeed See *misfeed.*

paper path The course of paper through the feeding, imaging/impression, and delivery sections of a copier or duplicator. See *delivery system; feeder; imaging system.*

paper tape One of the recording media used in editing typewriter systems. Normally codes are recorded at 10 to the inch at a cost of approximately 3 cents per 1,000 characters.

paper weight See *basis weight.*

paragraph indent Standard measure of indentation used to mark paragraphs or other indented material.

parallel printer A computer or word processing output peripheral which employs a parallel interface to connect an input terminal. The parallel interface allows the printer to accept transmission of data which are sent in parallel bit sequences.

parity bit A noninformation bit that is used to ensure that data have been transmitted accurately; a receiving device counts the 'on' bits of every arriving byte; if odd parity is specified, an error condition will be flagged any time an even number of 'on' bits are detected.

partial-payout lease See *operating lease.*

password A unique word or string of characters that a word processing operator must supply to meet security requirements before gaining access to the system. The password is confidential, as opposed to the user identification.

patch cord A wire or cable for connecting two pieces of audio equipment.

patching A method of transferring previously recorded material onto another medium.

pause control A control feature on some recorders that permits stopping the tape movement temporarily without switching from the "play" or "record" settings.

PAX   Private Automatic Exchange. A dial telephone exchange that provides private telephone service to an organization.

payout (or payoff) period   The time it takes for an investment (e.g., in a new machine) to pay for itself.

PBX   Private Branch Exchange. A manual exchange connected to the public telephone network on the user's premises and operated by an attendant supplied by the user.

PBX dictation system   A central dictation system using telephone-company wiring and dial or touch-tone telephones.

PC   See *photoconductor.*

perfector   An offset duplicator capable of printing on two sides of a sheet in one pass through the use of on-line, dual-impression systems.

performance-analysis budget   A variable budget composed of fixed and variable elements in which the segregation and the quantification of these two elements are determined by analyzing the probable requirements during the coming budget period under varying volume conditions (the cost-behavior characteristics).

performance variance   The difference between actual expenditures (or income) and the budget.

peripherals   Devices (such as printers, OCR readers, and communications) which may be configured with word processing systems as options, extending their capabilities. More sophisticated systems can frequently share a peripheral between multiple stations, making the use of a high-speed printer or other expensive piece of equipment cost-effective.

PERT   Program Evaluation and Review Technique. A planning technique and instrument of management control which uses network theory. A "network" is a plan of action of some contemplated activity, in the form of a logic chart or flow diagram, furnishing a graphic step-by-step picture of goals to be achieved, and their relationships and time estimates. This schematic enables the planner to see at a glance possible obstacles, relative priorities, progress of the work, etc. See *CPM.*

phonetic alphabet   An alphabet containing a separate character for each distinguishable speech sound.

photocell   See *photoelectric transducer.*

photocomposition   See *phototypesetting.*

photoconductor   A metallic substance such as selenium or cadmium sulfide which is capable of conducting and retaining electrical charges. If any portion of a photoconductor is exposed to light, that part will lose its charge. Photoconductive materials are employed in the electrostatic process to retain a latent image (charge) of a document, which is subsequently imbued with toner particles to create an image. See *electrostatic process.*

photoconductor belt   In plain-paper electrostatic copying, a photoreceptor-belt intermediate composed of a charge-sensitive organic material such as polyvinyl carbazol-trinitrofluorenone (PVK-TNF), or a flexible nickel base coated with selenium (e.g., Xerox 9200). The belt is flexible and is recirculated for repeated usage. This type of image transfer system is typical of high-speed copiers such as the Kodak Ektaprints 100 and 150, the IBM Copier II and Series III, and the Xerox 9200. See *electrostatic process; photoconductor.*

photoconductor drum   In plain-paper electrostatic copying, a photoreceptor intermediate composed of an aluminum core covered by a thin aluminum oxide layer which is coated with a charge-sensitive material, most commonly amorphous selenium (Se). Selenium drums are widely used in dry-toner imaging systems. In liquid-toner systems, a metallic drum core is coated first with photoconductive cadmium sulfide (CdS), then with a protective plastic lamination. Drums are employed in the image transfer systems of most plain-paper copiers. See *electrostatic process; photoconductor.*

photoconductor master   See *photoconductor.*

photodetector   See *photoelectric transducer.*

photodiode   See *photoelectric transducer.*

photodiode sensor array scanner   A type of scanner that employs a stationary configuration of tiny photodiodes arranged in a matrix that equals in width one lateral scan line of a document. An original is roller-fed from a flat-bed tray and passes by the sensor array one line at a time. Light is reflected from each line and focused through a lens onto the face of the photodiode array. Each diode acts as an independent photosensor in converting a small picture element into part of the total electrical signal for a scan line.

photo-direct exposure   Exposure of an original through a camera lens system directly onto the surface of sensitized paper, film, or plate material. This method allows for adjustments in the original image size. See *exposure.*

photoelectric reader   A light-sensing reader device used on tape-driven power typewriters.

photoelectric transducer   Component of a facsimile scanner that receives light and dark image patterns of the original and converts them into electrical signals for transmission over communications lines.

photographic printer   A high-resolution technique of facsimile printing that employs a focused light source to expose photopaper. The photopaper or film is wrapped around a cylinder in a light-sealed box. As the drum rotates, a glow modulator tube converts incoming electrical signals into a light source of variable intensity. The light source moves laterally across the length of the paper, creating one image line with each revolution of the cylinder. The light source varies in intensity with the strength of the electrical current, creating output that closely resembles the finer gradations of light and dark in the original.

photo offset　See *offset process.*

photosensitive　A term applied to a reprographics medium, such as silver-halide film or plates, that is capable of recording light images for extended duplication purposes. See *plate.*

phototypesetting　The setting of type via electronic or electromechanical optical systems onto photographic paper or film. Input to such devices can be by direct keyboard entry, paper tape, or magnetic media. Fonts are contained on reduced character matrixes through which light is flashed and focused to expose the photographic medium.

pica　In typesetting, a unit of measurement equal to 0.166 inch, which is about 1/6th of an inch, or 12 points exactly. See *point.*

pica type　A 10-pitch (10 characters per horizontal inch) typewriter setting. Also, a particular style of type face.

pilot model　A summarized combination of the flexible budgets of a company or a unit, which can be adjusted to incorporate anticipated or hypothetical situations, and which will then depict the manner in which costs would react and how the proposed changes would affect the profits, financial conditions, and cost–volume–profit relationships of the company.

pinbar　An offset plate cylinder clamp equipped with a row of pins to accommodate punched paper or metal plates. See *clamp; offset process; plate.*

pinfeed platen　A typewriter or printer platen that employs a sprocket-type pinfeed for the indexing of continuous-form paper.

pitch　Horizontal character spacing at 10 or 12 characters per inch. Ten-pitch spacing is called pica, 12-pitch spacing, elite.

plain paper　Economical, uncoated, base paper stock with a low rag content. Various grades of plain paper are available for use with offset duplicators and electrostatic copiers. Paper gradients will vary according to grain, basis weight, color, strength, stretch, and opacity. Plain paper for copiers generally requires a high moisture content (due to the heat involved in copying), whiteners for good image contrast, and some clay base to create a smooth finish. See *electrostatic process; plain-paper process.*

plain-paper copier　See *electrostatic process; plain-paper process.*

plain-paper process　An electrostatic copying process in which the image (charge) of the original is transferred via photoconductor intermediate (drum, belt or sheet master) to the surface of a plainpaper copy sheet. See *electrostatic process; photoconductor; plain paper.*

plate   In the offset process, a thin sheet of paper, plastic, or metal from which an inked image is transferred to the blanket cylinder for printing. Plate types include direct-impression, presensitized, wipeon, transfer, and electrostatic. See *direct-impression master; electrostatic plate; offset process; presensitized plate; transfer plate; wipe-on plate.*

platemaker   A device that exposes presensitized plate material for use on offset duplicators. Platemaker types may be classified as follows: vacuum frame exposure-only units, necessitating off-line processing by hand or machine; photo-direct camera platemakers requiring off-line processing by hand or machine; photo-direct automatic camera/platemakers with on-line processing of plate material. See *automatic camera/platemaker; plate.*

platen   See *exposure platen.*

platen cover   The rigid or flexible cover on the exposure platen of a copier. See *exposure platen.*

playback   Includes all console preparation, automatic playback control handling, and machine playback time.

playback mode   The amount of information played back from the recording.

playback print rate   The automatic typing speed of the printer. It should be noted that these are maximum speeds for best situations. In actual practice, printers operate at somewhat slower speeds, depending on the amount of white space (as in tabular work or short lines), carriage returns, unidirectional or bidirectional printing capabilities, and other factors.

PL/I   A high-level programming language, designed for use in a wide range of commercial and scientific computer applications, which has features of FORTRAN and COBOL plus others.

plug-in counter   A device used to monitor copier usage. The system consists of a terminal mounted on a copier which serves as a receptacle for individual counters that are "plugged in" to activate the equipment. Different counters are distributed to different departments or employees within a company. At the end of a billing period, the counters are collected and the figures tallied to determine the amount of copier usage per counter-user. See *copier monitor device.*

PM   See *preventive maintenance.*

point   A unit of type measurement equal to .0138 inch or approximately 1/72 of an inch. There are 12 points to a pica. See *pica.*

point size   The vertical space allocated to a type face, referred to in the horizontal value of measurement called points. For example, there are 8-point, 10-point, and 15-point type sizes. See *point.*

point-size range The range of operator-selectable type sizes as generated by a photocomposer, which is expressed in point sizes from minimum to maximum. Type size is a function of the photocomposer's optics, which work in conjunction with the font master to either magnify or reduce the size of type required. See *font master; point size.*

point-to-point A limited network configuration with communication between two terminal points only, as opposed to multipoint and multidrop.

polarity Occasionally the tones generated by a specific telephone do not control the recorder. This means that the sounds are not in the same polarity as the tone recorder. When this occurs, the telephone company has not put a "polarity guard" on that specific telephone. This feature must be on all telephones that are to be used with the touch-tone dictation system.

polling A communications feature that allows one or more stations of a communicating word processing system to check with other systems to see if a message is ready to be sent.

port An input/output channel including the physical connector and control logic to interface a peripheral device with a mainframe.

portable dictation unit A dictation unit designed for in-the-field use. Portable dictation units vary in weight from about one-half pound to two pounds, and range in size from small enough to fit in a shirt pocket or purse to as big as a cigar box. Recordings are made on magnetic or plastic belts, or on magnetic-tape reels, disks, cassettes, or microcassettes. Each of these media is easily mailed or forwarded back to a central office for transcription on a compatible desktop unit.

positive corotron See *corotron.*

power keyboard See *magnetic keyboard.*

power typing An application or system that employs word processing equipment. Generally used to describe a low-level application such as typing repetitive letters.

PPC Plain-Paper Copier. See *plain paper; plain-paper process.*

pre-clean corotron See *corotron.*

predetermined time Standard times for routine clerical operations, derived by time studies.

pre-mix See *toner pre-mix.*

preparation Includes handling paper, loading the recording media, console preparation, and typewriter preparation (tabs, margins, margin resets).

prerecorded Text stored on magnetic media for subsequent playout as part of a repetitive letter or a letter created from boilerplate. Variable information, either prerecorded or keyboarded, may be combined with such prerecorded text.

presensitized plate   Plate material that has been precoated with light-sensitive material. Types of plate coatings include silver halide, diazo compounds, photoploymers, zinc oxide, and thermographic. See *coated-paper process; diazo process; plate; thermographic process.*

pressure diazo   In diazo duplicating, an AM/Bruning term for the activator development process. See *activator development; diazo process.*

preventive maintenance   Upkeep operations necessary to keep equipment running in good order, including regular cleaning, replenishment of expendable supplies, and the replacement of worn or expendable parts. Some of these functions are usually handled by manufacturer service representatives in accordance with a general equipment-maintenance contract.

principal   A person who originates work; also called an "originator" or "word originator."

printer   The unit that types out the page, frequently an IBM Selectric II Typewriter.

printer types   The most commonly used printers in WP are the IBM Selectric II typewriter and the Qume or Diablo daisywheel printer. By far the largest number of installed word processors continue to employ the 15.5 cps Selectric typewriter, either in the form of ordinary off-the-shelf models or heavier-duty models designed for use in word processors. Such specially engineered units offer more durability and reliability. A few systems employ the IBM Correcting Selectric and IBM Executive typewriter in word processing systems.

An increasing number of systems (the vast majority of all non-IBM equipment) employ the new daisywheel mechanism, which offers higher output speeds (30, 40, 45, or 55 characters per second) and a variety of type styles.

In 1976, IBM introduced an ink-jet printer, now available as both a stand-alone adjunct to mag card word processing systems and as the output mechanism for the new IBM Office Systems 6 (OS 6) series. This printer spits out a stream of tiny, electrostatically charged ink drops at the page, placing them precisely as its head moves across the paper. Print speeds range from 77 to 92 cps, and output considerations allow for a variety of formats, type styles, and type sizes. Unique to this system is the ability to automatically access a choice of two types of paper plus envelopes. While there are limitations to the ink-jet printer (no carbon copies, paper limitations, no masters), the ink-jet unit has been received with enthusiasm and is likely to be imitated by other vendors.

print folder   A device used to automate the folding of large reprographic prints, such as diazo copies of engineering drawings. Such equipment may be operated off-line or on-line with the delivery system of a duplicator.

printout   See *hard copy.*

printout queuing   A feature which allows a number of documents to be lined up or queued for subsequent printout while the operator goes on to perform other tasks. Such printout queuing may be quite primitive, employing only a single printer and handling one page or one document at a time. Other queuing may be very sophisticated, with multiple printers and print queues for each printer, capable of processing large documents or allowing multiple documents in the queue. Some systems also allow documents to be deleted from the queue and/or priority documents to be processed ahead of a normal first-in, first-out queue.

print suppression   Control codes entered during text entry/editing such that designated portions of text will not print during playback.

printwheel   See *daisywheel printer; element printer.*

private wire   A dictation recorder system that does not use telephone-company lines.

processor   A computer or part of a computer capable of receiving data, manipulating them, and supplying results.

producer   A person who, in word processing, types the document.

product cost   The summation of the costs required to produce a product, recognizing that the product may have been worked on in several departments and in conjunction with other products.

production   A measure of the goods or services generated by work. It includes both quantitative and qualitative aspects, such as number of clients served or percentage of defective units produced. Less tangible aspects of production would include customer satisfaction or ratings of workmanship.

productivity   Output per man-hour.

program   A set of instructions arranged for instructing a word processor or computer to perform a desired operation.

program, operating or budget   A fiscal expression of the operating intentions of a company or unit for a budget period as they are anticipated at a given date. This is arrived at by applying the volume forecast to the budget formula in every component area to indicate the planned income, costs, and profit under the stated conditions.

PROM   Programmable Read-Only Memory. Refers to the solid-state memory for storing programs which a vendor company can program to customize a system before delivery to the user. Generally, PROM cannot be altered once it has been programmed.

prompt   Reminder(s) usually implemented via display that assists the word processing operator in performing a function.

proofreading   Reading copy to detect and mark errors to be corrected.

proportional spacing   Typed, printed, or displayed text where each alphanumeric character is given a weighted amount of space. For instance, an "I" might be two units wide, an "L" four units wide, and a "W" five units wide. Such output has a printlike appearance, especially when combined with a character-spacing scheme employing sophisticated intercharacter spacing.

protocol   A formal set of conversions governing the orderly exchange of information between communicating devices by defining such things as connection establishment, security provision, data sequencing, error control, etc. Protocols achieve efficient line utilization by reducing the amount of information transferred by distinguishing between device control information and data.

punched paper tape   A roll of paper on which holes representing characters are perforated and then read back to produce documents.

put-away   Cleaning up after a task is completed.

PVK-TNF   Polyvinyl carbzol-trinitrofluorenone. See *photoconductor belt.*

quality control   A check on work to keep a uniform quality and appearance, always striving to standardize where possible.

quality control, statistical   Control of product quality by sampling techniques.

quality printer   See *letter-quality printer.*

queue   Storage areas within a computer-based word processing system, with each item in storage linked to the items before and after it to form a queue or line. Queues are usually created to permit an individual operator to send text to be processed (for such functions as data communications or printout) and to then continue to perform other work rather than waiting for access to the appropriate function or peripheral.

queuing problem   The problem of determining the facilities to be provided when the need for them varies at random—for example, the optimum number of ticket windows in cases where there may be long queues at some times and no customers at all at other times.

QWERTY keyboard   A standard typewriter alphanumeric keyset, as carried over from the printing industry, named for the first six keys of the third row from the bottom.

ragged left/right   Refers to ragged, or uneven, nonjustified, right or left margins.

RAM   Random-Access Memory. Storage or memory that allows data (such as documents) to be stored randomly and retrieved directly by an address location. The system accesses the addressed material, with no need to read through intervening data. Information may be retrieved more speedily from random access memory than from serial media such as tape.

reader   A special machine that enlarges microfilm or microfiche so that the viewer can read the miniaturized information.

read/write head   The mechanism which writes data to or reads data from a magnetic recording medium.

real-time operation   Term used of a computer that is processing data fast enough for it to be used to control the operation that is supplying the data.

ream   Five hundred sheets of paper; a standard of paper packaging.

recall   Bring information from storage into memory for typing out.

recirculating ammonia development   In diazo duplicating, an ammonia development system which recycles unused vapor back into the main supply source. Anhydrous or aqueous ammonia may be employed. See *ammonia development; diazo process.*

recirculating feeder   An automatic document feeder that refeeds original documents or document sets for multiple, repeat exposures. Such systems are employed on some copiers and diazo duplicators.

record   A collection of related items of data (fields) treated as a unit.

recorder   A component of a dictation system that records the dictation onto magnetic media.

records processing   Refers to the manipulation of files of information, such as selecting from certain fields, and resequencing files into different categories; and the generation of reports from fields of data.

reduction   The ability of a copier or camera system to reduce the image of an original document through the use of a lens with adjustable focal length. In copying, this permits the reproduction of a large document original onto a standard sheet size.

reel-to-reel   Refers to tape recorders that do not use a cassette or cartridge.

reference code   An electronically recorded magnetic-tape indexing point.

register   A special section of main memory where data are held while being worked on.

reliability   The extent to which equipment or systems will operate as necessary to achieve the purpose for which they are intended.

remote access   Pertaining to communication with a computer processor by terminal stations that are distant from that processor.

remote batch   A method of entering jobs into the computer from a remote terminal for processing later in a batch processing mode. In this mode, a plant or office geographically distant from the central computer can load in a batch of transactions, transmit them to the computer, and receive the results by mail or via direct transmission to a printer or other output device at the remote site.

rent-purchase plan   A rental arrangement in which a certain percentage of monthly rental payments can be applied to the eventual purchase of the equipment. See *conditional sale.*

repagination See *automatic repagination.*

repeating key A typewriter key which continues typing (or recording) as long as it is depressed. Also called a typomatic key.

repetitive typing Documents, such as form letters, that are typed over and over again, usually for mass distribution.

replenishment kit An equipment supplies package that is good for a specific volume of reprographics work. Copier supply kits, containing toner, developer, cleaning brushes, or other materials, are often sold in this manner.

reprographics Refers to the reproduction and duplication of documents, written materials, drawings, designs by photocopy, offset printing, microfiling, office duplicating, etc. Also includes all auxiliary bindery operations.

residual value The value of a piece of equipment at the end of a lease term.

response time The time a system requires to respond to an operator command in supplying stored data or completing a process cycle.

restart In word processing, to begin a typing job over again.

reverse leading The ability of some phototypesetters to allow the reverse movement of the photographic medium once material has been typeset. This permits the setting of side-by-side columns on the actual composed page as well as other special effects. In practice, the operator may compose one complete left column, then through reverse leading (pronounced *ledding*) back up the photopaper in order to compose a right-hand column alongside the left column.

reverse mode An operational feature of some diazo duplicators which permits the reversing of the exposure feed to prevent misfed or misaligned originals from being damaged. See *diazo process.*

reverse search See *direct reverse search.*

revision and status reporting The ability of a word processor to automatically create an index, listing the documents stored on a tape, diskette, or disk. In some cases, the index may be used as a status report, and may show such entries as the date on which the document was created, the last date on which the document was accessed and/or revised, the number of revision accesses which have occurred, and so forth. Some systems can sort indexes by author, department, operator, etc., and print a status report.

revision cycle Path of a typed document from initial keyboarding to final output.

RJE Remote Job Entry. Input of a batch job from a remote site and a receipt of the output via a line printer or card punch at a remote site. The technique allows various systems to share the resources of a batch-oriented computer by giving the user access to centrally located data files and access to the power necessary to process those files.

RMN   Remote Microphone Network.

RO   (1) Receive Only. A communicating device that operates in a receive mode only. (2) Receipt of Order.

robot typer   A unit that usually operates an electric typewriter from a "piano roll" device.

roll-fed   See *roll feeder*.

roll feeder   A paper feeder on a copier that uses roll paper supply and console-selectable copy-length control.

roller train   A series of rollers employed on offset duplicators to carry fountain solution and ink to the plate cylinder.

ROM   Read-Only Memory, a solid-state memory for programs that is inflexible and cannot be altered.

rotary feeder   See *single-sheet document feeder*.

rotating-cylinder scanner   Common type of facsimile transmitter in which the original is mounted around a cylindrical drum and scanned by an optics/photocell assembly (scan head) parallel to the length of the cylinder. The drum rotates and the scan head moves across the document, scanning one line width per revolution.

rotating helical aperture scanner   A type of facsimile transmitter fairly common on low- to mid-volume fax units. The original is roller-fed over a flat-bed copy platen and illuminated by an area lamp. Lens and mirror optics reflect and focus one scan line of the moving document at a time, first through a fixed horizontal slit aperture and then through a rotating helical aperture. The rotating of the helical aperture produces one lateral scan of the original per revolution, with the passing light being focused a finer time onto a photocell for conversion to an electrical signal.

rotating lens turret scanner   A facsimile transmitter employing a scan head composed of a stationary mirror surrounded by a rotating turret of lenses. The original is mounted on a moving semicylinder platen that feeds the document as it is scanned one line at a time. While rotating, each lens focuses scan lines onto the mirror, which in turn projects the light through an aperture and onto a photocell.

rotating scan head   A facsimile transmitter employing an optics/photocell assembly that rotates as it scans each line. The document is mounted on a semicylinder platen which moves beneath the scan head.

RPQ   Request for Price Quotation.

RS-232   A technical specification published by the Electronics Industries Association that establishes interface requirements between word processors, modems, computers, and communications lines.

run-length code A method of redundancy reduction (data compression) utilized by digital facsimile transmitters to enhance speed. When image patterns of an original are converted into digital signals, all black and white areas on a page are reported as a series of ones (black) and zeros (white). The number of white spaces between black elements (number of zeros between ones) is assigned a number, or run-length code. Each time a space is encountered, the unit assigns a short code to represent it rather than reporting all of the individual white spaces (zeros), and the most frequently used run-length codes are given the shortest binary numbers. Run-length coding may be performed horizontally across the width of the page (one-dimensional system) or vertically as well (two-dimensional system).

scanning In dictation terms, refers to rapid listening and identification of specific parts of a recording.

scorer A device that uses a rule or string to imprint or impress a sheet of paper to accommodate folding.

scratch pad memory An internal storage area reserved as an intermediate working area.

screen See *halftone screen.*

scrolling See *horizontal scrolling; page scrolling; vertical scrolling.*

SDLC Synchronous Data Link Control. An IBM communications line protocol. In contrast with BSC, another IBM communications protocol, SDLC is more efficient and provides for full duplex transmission.

Se Selenium. A possible coating for a photoconductor, commonly used on drums and belts in plain-paper, dry-toner systems. See *photoconductor; plain-paper process.*

SE Systems Engineer. A vendor representative who provides engineering or programming support.

sealed system The electronic text-editing typewriter counting process to locate referenced material on a tape.

search capability The method employed by a word processor to locate an editing point. Unsophisticated systems usually search for document, page, or paragraph number (reference code), or by line number. More sophisticated systems can also search by character string, having the ability to access the occurrences of some set number of sequential characters.

second-generation typesetter An electromechanical typesetting device which employs a font master, lens, and flash exposure system to set type onto photographic media. This classification includes both stand-alone typesetters and direct-entry devices. See *font master.*

second sheet   A thin sheet of paper used for carbon copies of the top page.

security   See *system security.*

seize   To electronically connect with a dictation recorder.

select   See *automatic file select.*

Selectric   See *printer types.*

self-logging   A worker's record of time required to do a job.

separation   In the diazo duplicating process, the separation of the original and copy paper, either manually or automatically, after exposure and just prior to development. See *diazo process.*

sepia line   A brown-colored print line available with various forms of diazo-coated material. See *diazo process.*

sequential access   Data or storage, such as magnetic tape, that must be searched serially from the beginning to find any desired record.

serial access   See *sequential access.*

serial printer   A computer or word processing output peripheral which prints out successive characters one at a time, either from left to right or bidirectionally for increased speed. Also refers to the interface used to connect the printer to the input device. The interface allows the printer to accept transmission of data which are sent serially, or one character at a time.

serial storage   A storage media organization in which data or text are serially recorded one character or text block after another. Text access points are retrieved by serially searching through the medium (usually a magnetic-tape cassette or cartridge). Such storage is adequate for most word processing unless it is necessary to revise long documents or to move quickly from one access point to another. In such instances, random-access storage media provide faster access.

set-off   See *offset.*

setup time   Time spent preparing to do a job.

shared-logic word processing system   A multiterminal (operator console) system where each terminal shares the word processing power, storage, and peripherals of a central computer. Included in this category are distributed logic systems which share peripherals and sometimes storage but have most computing power (logic) distributed at the individual operator stations.

shared-resource system   See *shared system.*

shared system   A multiterminal word processing system configuration whereby a number of intelligent, independent workstations share common peripheral devices such as printers, OCR readers, disk files, etc.

sheet-fed   See *sheet feeder.*

sheet feeder   A paper-feed mechanism on a copier or duplicator that uses a precut sheet paper supply.

sheet master   See *master; photoconductor.*

shield   A piece of plastic or cardboard used to prevent smudges when erasing on original copies.

shift key   A typewriter key that selects the top-marked symbol on a character key.

simultaneous printout   The ability of the sytem to print one document while simultaneously recording new text or revising previously recorded text. Some systems accept one page or one document at a time for simultaneous (also called background) printing; others queue the documents to be printed. If the system creates a print queue, printout may occur on a first-in, first-out basis, or the operator may be allowed to make priority assignments.

single-sheet document feeder   A document feed mechanism on a copier that allows the mechanical carrying of originals, fed manually one at a time, across the exposure platen. The device is positioned over the platen and usually utilizes rubber rollers and guide belts to transport the original. The use of a single-sheet document feeder increases throughput time by eliminating the manual step of positioning each document. See *automatic document feeder.*

skip capability   The ability of the system to skip over certain segments of text at operator command to edit a document on paper or on display while leaving it unchanged on the magnetic medium. May also include the ability of the operator to record comments between special skip codes so that the comments will not be printed during the final output, but are available for reference either on the display or on draft copies.

slave   I/O or print-driven module driven by a master unit. In some low-level word processing applications, it is common to have a master unit plus a number of slaves, automatically grinding out repetitive letters.

slit exposure system   A form of copier exposure system in which the exposure illumination and lens assembly are positioned under a narrow slit over which a moving platen carries the original. See *electrostatic process.*

slitter   A device that cuts printed sheets into two or more sections via cutting wheels. A slitter may be attached to a press or folder, and may handle precut sheets or webs.

SNA   Systems Network Architecture. IBM's standardized relationship between its virtual telecommunication access method (VTAM) and the network control program.

snapout carbon pack ("copyset")   A pad of thin paper and carbon sheets in which a carbon is attached to the bottom of each sheet of paper.

soft hyphen   See *discretionary hyphen.*

soft-sectored A term used to describe a particular diskette format, and way of recording on the diskette. Soft-sectored disks are preformatted, having data fields that are changed and updated. The first track of a soft-sectored diskette identifies the disk; the next four tracks store basic format information such as the track and sector location of stored material. Since soft-sectored diskettes require a format, less storage capacity is available (250K bytes out of a possible 400K) for text storage than is available with hard-sectored diskettes.

software A term coined to contrast computer programs with the "iron" or hardware of a computer system. Software is a stored set of instructions which govern the operation of a computer system and make the hardware run. Also used in word processing to mean all of the non-hardware parts of the system, including manuals, training, etc.

software-based Refers to a word processor system design whereby word processing (or other) software is "loaded" or read into the system's random-access memory via a form of media upon system startup, as opposed to software that is resident in read-only memory.

software house A company which offers software programs and support service to users. This support can range from simply supplying manuals and other information to a complete counseling and part-time computer programming service.

software-programmable A word processing system whose functions are defined by a program, generally supplied by the manufacturer, that may be redefined or updated by changing or replacing the program.

solid-state device Any element that can control current without moving parts, heated filaments, or vacuum gaps.

sonic search Fast-forward scanning of voice to locate a special spot of dictation.

sort See *automatic file sort.*

sorter A device that collates multiple copies into groups or sequenced copy sets, usually operated on-line with a copier or duplicator. The term "sorter" generally applies to a high-volume application. See *collator.*

sound-on-sound A method by which previously recorded material may be rerecorded while simultaneously recording new material.

sound sheets A flat, magnetic vinyl medium for recording dictation.

speakerphone A speaker on a desk through which voices are transmitted both by the person calling from the office and the person calling in from outside.

speed (1) The rate at which a reprographics device produces duplicates. (2) In diazo duplicating, the light-sensitivity of the copy material, which is directly related to the required exposure time. (3) The light-sensitivity of any photographic material, relative to exposure time.

spiral comb binder   A type of office binding device that punches a series of holes along the left edge of the document to be bound and fastens the pages together via a spiral plastic comb that is threaded through the holes. See *binder.*

spirit duplicator   See *spirit process.*

spirit master   An image master employed in the spirit duplicating process. Carbon sheets are used to produce spirit master images. Common master colors are purple, black, blue, red, and green. Multiple carbons may be applied to a single master to produce multiple colors. Masters can be made via direct-impression or thermal copiers. See *spirit process; thermographic process.*

spirit process   A duplicating process that uses a negative image master and moistening solution to produce copies. The solution (usually methyl alcohol) is used to moisten blank sheets of paper before they are pressed into contact with the master. The moistened sheet surface then dissolves and absorbs a small portion of the carbon dye on the back of the master to form a positive print on a sheet of paper. All images are formed by taking dye from the master and transferring it to paper—no separate inks are involved. Various colors may be used on a single master by employing different color carbon impression sheets. Also called *fluid duplicating.* See *spirit master.*

split keyboarding   Keyboarding and editing on one unit and playing back on another.

squeal   Noise caused by sticking tape.

stand-alone word processing system   The classic, single-station word processor such as a mag keyboard or video display system which does not share the processing power of a central computer. See *shared-logic word processing system.*

standard data   A predetermined time standard obtained and accepted for a particular activity as reliable and representative of the work performed.

standard hour   The standard amount of work to be done in an hour.

standardization   Using one type of component, material, or machine for as many purposes as possible.

start up; startup   To begin an operation; the beginning of an operation.

stat typing   Priority typing.

statement size   The common sheet size for billing statements, which is $5\frac{1}{2}'' \times 8\frac{1}{2}''$.

station   The record/playback drive mechanism on an editing typewriter; e.g., the MT/ST Model 4 has two stations.

stencil duplicator   See *stencil process.*

stencil master   An image master employed in the stencil duplicating process. It consists of a backing of fibrous-base tissue material that is permeable by ink, and an overcoating of plastic that is ink-impervious. Displacing the plastic coating leaves a continuous stencil on an image supported by fibrous, ink-transmitting material. Stencil masters may be made by direct impression, in thermal copiers, or by electronic stencil cutters. See *stencil process.*

stencil process   A duplicating process that uses stencil masters and a machine inking system to generate reproductions. After preparation, a stencil master is mounted on a cylinder containing a fabric inking pad, or on an inking screen belt supported by two inking cylinders (single-cylinder and dual-cylinder stencil duplicators). Ink is passed from the pad or belt through the stencil and onto the paper surface as sheets are pressed against the master. Also called mimeograph duplicating. See *stencil master.*

stenotype   A silent shorthand machine keyboarding system to rapidly record speeches, minutes, etc., at speeds up to 400 wpm. Uses 23 keys and numeral bar printing in all caps, on a continuous paper roll. Entire words can be keyboarded with a single stroke.

stop code   A reference code recorded on magnetic media which causes the system to stop during printout. Used to allow the operator to perform such manual procedures as changing fonts or paper on the printer.

storable   The portion of a prerecorded document that is not subject to change. Examples are the format of a will or the body of form letters.

storage capacity   Total amount of text stored per unit of media (card, cassette, diskette) which may be accessed by the system without changing media. A magnetic-card system has about 5,000 to 10,000 characters "on-line." A magnetic-tape cassette would hold up to 300,000 characters. Diskettes hold about 250,000 to 300,000 characters, but in many cases the word processor's operating-system software is also stored on the diskette, so only 60 to 100 pages or so of storage are available. Disks can hold much larger quantities of data, frequently 2 million to 50 million characters or more on each disk.

store-and-forward   The handling of messages or packets in a network by accepting the messages or packets completely into storage, then sending them forward to the next center.

stored-form recall/display   The ability of a word processor to store a form (such as text, scale, lines, or a combination of them) and display it upon demand. The operator can then combine the form with new keyboarded text, then print out the completed form, and/or store the form with text or the form and text separately.

stored multiple formats   The ability to store more than one format, and to access the desired format on operator command. Some systems have a formal routine for this operation. Other systems do not, but can allow the operator to store a format statement on a special segment(s) of media and access it at will.

strike-on The process of setting type through direct impression via a device such as a composing typewriter. No photographic medium is involved; material is typed directly onto the paper that is used to prepare page layout.

strikeover One character typed over another.

stroke counter A device fitted on the typewriter that counts every key depression that is made. While this method is useful as a supplement to other methods, complete reliance on it demands very strict staff supervision to avoid erroneous records.

sub/superscript printout The ability of the printer to print characters a fractional increment (sometimes adjustable) above and below the line for footnotes, formulas, etc.

supervisor's console Usually refers to a panel from which tank recorder dictation can be distributed manually or automatically.

supplier See *vendor*.

support requirements Secretarial assistance required by a principal.

switch code A reference code recorded on magnetic media which instructs the word processor to alternate between media stations, allowing the combination of material from two sources, e.g., a standard letter and a name-and-address list.

switched line A telephone line that is connected to the switched telephone network.

switched network A multipoint network with circuit-switching capabilities. The telephone network is a switched network, as are telex and TWX.

switching center A location where an incoming call/message is automatically or manually directed to one or more outgoing circuits.

synchronous A mode of data transmission in which data are transmitted in blocks of characters. Sync characters are transmitted prior to actual data transmission to establish timing. Commonly used for batch transmission.

system security A number of systems provide a key lock, generally in place of the on/off switch, to prevent unauthorized access to the word processing system. A few of the more sophisticated systems require the keying of confidential password codes before the system itself, or specific documents, are available for access or revision.

systems typewriter A typewriter with some internal logic to perform special functions.

system survey The study, analysis, and improvement of the systems that service, control, and coordinate all the operations of an enterprise.

tab   Can refer to a central recorder indexing slip, or indentation space.

tank recorder   See *endless-loop recorder.*

task data sheet   A record of jobs by time period during a single work day.

task list   A detailed record of each type of work performed by each worker and the average number of hours spent to do it per week.

telecommunication lines   Telephone and other communication lines that are used to transmit messages from one location to another.

telecommunications   The transmission/reception between terminals, or between terminals and computers, of digitized information over telephone lines.

teleconference   A meeting of geographically separated conferees connected simultaneously via a telecommunications system utilizing two-way voice and/or video message communication.

telecopier   A facsimile device.

teleprinter   Equipment used in a printing telegraph system. See also *teletypewriter.*

teleprocessing   The processing of data that are received from or sent to remote locations by way of telecommunication lines. Such systems are essential to hook up remote terminals or connect geographically separated computers. See also *telecommunications.*

Teletype   Trademark of Teletype Corporation; usually referring to a series of different types of teleprinter equipment such as tape punches, reperforators, and page printers, utilized for communications systems.

teletypewriter   A generic term referring to the basic terminal equipment made by Teletype Corporation, and to teleprinter equipment. The teleprinter KSR (Keyboard Send/Receive) receives the line signal and prints the same as an RO (Receive Only), but in addition it has a keyboard that is used for manually sending line signals. It has no paper-tape capability but is very popular for conversational time-sharing and inquiry–response applications. The teletypewriter ASR (Automatic Send/Receive) combines the other devices into one machine containing keyboard, page printer, paper-tape transmitter, and paper-tape punch. Paper tape can be prepared off-line, and this can take place while hard copy is being received from the line or while other paper tape is being transmitted.

telex   An automatic teleprinter exchange service available worldwide through various common carriers; in the United States, Western Union is the carrier.

temporary margins   The ability to set a second, different margin for such purposes as an indented paragraph or quotations.

ten-pitch A typewriter spacing of ten characters to the horizontal inch; also known as pica spacing.

term dictionary/glossary The ability of the system to store technical vocabulary or frequently used phrases (such as the corporate address, signature blocks, etc.) and retrieve them via a significantly smaller number of alphanumeric keystrokes (usually two). This feature can greatly speed text entry for legal, medical, engineering, and other applications with complex vocabulary or a large number of boilerplate phrases. On systems which do not have this feature but do have a global search and replacement routine, it is possible to designate any word or phrase with an "impossible" combination (say "gx"), and then to replace the combination throughout the document before storage or printout.

terminal In general, a device that is equipped with a keyboard and is connected to a computer or word processor for the input of text/data.

text editing A general term that encompasses any rearrangement or change performed upon textual material, such as adding, deleting, or reformatting.

text editor A typewriter that permits stored information to be edited before printout.

Textverarbeitung The original term meaning "word processing" coined in Germany in 1965 by IBM.

therblig The smallest discrete human motion that can be identified in the performance of a task.

thermal copier See *thermographic process.*

thermal master Spirit or stencil master produced by a thermal copier. See *spirit master; stencil master; thermographic process.*

thermal printer A nonimpact printing technique which utilizes a special heat-sensitive paper. The paper passes over a matrix of dot heating elements. As data are fed to the printer, the dot elements relating to specific characters are heated, which changes the color of the paper at that point to reveal individual characters.

thermal tape binder A type of binding device that employs a fabric and glue tape strip applied along the spine of a multiple-page document, then pressure heated to create a bond. See *binder.*

thermographic process A copying process which employs heat-sensitive paper. A sheet of copy paper is placed on top of an original, and both are exposed to infrared light. When infrared light strikes the image or dark portion of the original, this area absorbs the light and increases in temperature. The increased temperature of the original image darkens the copy paper in a corresponding area, creating a reproduction of the original. The process is sensitive only to carbon-based (pencil; electrostatic copies) or metallic-based original impressions; most ballpoint inks will not reproduce. The infrared light is also insensitive to certain colors. A thermal copier may be used to make transparencies and spirit or stencil masters and to perform plastic laminations.

thermography See *thermographic process.*

third-generation typesetter An electronic typesetting device which uses a laser or CRT mechanism as the source of exposure for the photographic medium. Fonts are stored digitally; therefore, these devices do not employ font masters and lens systems.

third-party lease A leasing arrangement in which an independent firm buys the equipment from the manufacturer and in turn leases it to the end user. The "middleman" firm is known as the third party.

throughput The total of useful information processed or communicated over a given period of time.

tie line A private-line communications channel provided by communications common carriers for linking two or more points.

time-shared services Commercially available access to a computer, on a time and storage charge basis, allowing the user to connect a communicating word processor or terminal to the time-shared service's system. Employed by infrequent users who have sophisticated needs as well as others who wish to access special data bases such as specifications libraries.

time sharing The sharing of the power (and cost) of a large computer facility among a number of users, each equipped with terminals.

time study A method of determining the time necessary to perform a task. Procedure may involve timing the various elements (e.g., very small components of the task such as picking up material, positioning it, or inserting it in the machine) with a stopwatch, totaling the times, and adding "allowances" for such things as fatigue and personal time.

tone control A feature used to vary treble and bass response during playback.

toner In the electrostatic process, minute dry particles of resin and carbon black that are used to create images. Toner is capable of accepting an electrical charge. It is carried to the photoconductor by a developer medium and transferred to the surface of a copy sheet by a series of successively greater electrical charges. See *electrostatic process.*

toner concentrate Pure toner in concentrated form which must be added to the developer medium of a copier on a regular basis to maintain balanced copy quality. Also known as toner intensifier. See *toner.*

toner intensifier See *toner concentrate.*

toner pre-mix   A toner and liquid developer mixture that is packaged and ready to use. See *developer; toner.*

toner reservoir   A refillable toner supply container on a copier which feeds toner powder to the imaging mechanism. See *electrostatic process; toner.*

touch typing   The skill of typewriting without looking at the keys.

track   The path on magnetic tape or card along which a single channel of sound or codes is recorded.

trailing end clamp   The clamp on a plate cylinder that secures the bottom edge of an offset plate to the surface of the cylinder.

transcription   Making a written copy of dictated or recorded matter in longhand or on a typewriter.

transcriptionist   A person who types (transcribes) recorded dictation into document form.

transfer   (1) In the plain-paper copying process, the transference of toner particles from the photoconductor intermediate to the copy sheet. See *electrostatic process; toner.* (2) In word processing, to copy material from one medium to another.

transfer corotron   See *corotron.*

transfer plate   A platemaking procedure in which a light-sensitive intermediate material is used to form an image, which is transferred or printed onto a plate. Original copy is either contact- or projection-printed onto a negative material treated with a silver light-sensitive coating, producing a latent image. Successive processing steps include: sandwiching of the negative with an aluminum plate; immersion in a developing bath which converts the silver particles to black metallic silver and transfers the image by diffusion to the plate surface; separation of the negative and plate; fixing of the plate with a lacquer to clarify the image. Transfer plates are used to duplicate typed material, line drawings, and occasional halftones. See *plate.*

transmit   To send data from one location and receive it at another location.

transparency   A clear or tinted plastic or acetate sheet containing a printed image. Transparencies may be made on electrostatic and thermal copies. See *electrostatic process; thermographic process.*

triboelectric property   The inherent electrical charge of copier developer material, usually positive. See *developer.*

true lease   A type of *financial lease* under which ownership of the equipment remains with the lessor. To qualify as a true lease for tax purposes, the Internal Revenue Service Ruling 55-540 of 1955 states that: (1) title must remain with the lessor; (2) the rental payment must be competitive with industry rates, represent payment for use, and have a rate that does not vary appreciably with or without purchase option; (3) the option-to-purchase price must not be less than the fair market price at the lease's expiration date; and (4) equity cannot be allowed on rental payments. For tax purposes, total monthly payments can be deducted.

trunk   A major link in a communication system, usually between two switching centers.

tuning   Magnetic belt dictation transcriber term referring to alignment of transcription head to correspond with the track made by the recorder dictation head.

turnaround time   The elapsed time between the submission of a word processing job to an operator and the return of the results.

turnkey system   A complete system configuration of integrated electronic components, including a processor, I/O devices, software, and programming, which is delivered and is fully operational upon plugging into a power source. The system is supplied from a single source; however, several vendors' products may make up the total system, with responsibility being divided among the vendors.

twelve-pitch   Refers to 12 typewritten characters to the horizontal inch; also known as elite spacing.

two-sided copying   See *duplex* (for copying); *perfector* (for duplicating).

TWX   Teletypewriter Exchange Service. A public teletypewriter exchange (switched) service in the United States and Canada formerly owned by AT&T but now belonging to Western Union. Both Baudot and ASCII-coded machines are used.

typomatic key   See *repeating key*.

typebar   Conventional typewriter mechanism.

typing application   A definable unit of paperwork requiring typing.

typing stations   Points where typewritten work is done throughout a plant; number of units in a WP center.

UL   Underwriters' Laboratory.

ultrafiche   Microfiche with images reduced more than 90 times.

ultraviolet lamp   The type of lamp employed in the exposure section of a diazo duplicator. The ultraviolet source may be either a mercury vapor lamp or a fluorescent type ("blacklight"). See *diazo process*.

unaffected correspondence   Correspondence expressed in hours of work not to be completed on magnetic keyboard equipment. These hours are considered administrative hours.

unattended operation   A mode of operation of a word processing system whereby because of the automatic features of the system certain functions, such as communications and printing, may be performed without an operator in attendance.

unattended recording   A recording unit set up so that anyone can dictate without the need to have someone change recording media—the recorder is always ready to receive dictation.

unbundled   The services, programs, training, etc. which are sold independently of the system hardware by the manufacturer. Thus, a manufacturer who does include all products and services in a single price is said to be "bundled."

underline   See *automatic input underline.*

underscore display   Underscored characters may be directly displayed, or may be indicated by display codes placed before and after material to be underscored on printout.

unit sets   Preassembled packets with interleaved one-time carbons, or carbonless paper.

unloading   The act of rewinding a magnetic tape and removing it from the editing typewriter.

upper case   Capital or large letters. Some draft or line printers print in upper case only rather than in upper and lower case as with a typewriter.

USASCII   United States of America Standard Code for Information Exchange.

USASI   United States of America Standards Institute.

user group   Any organization made up of word processing users (as opposed to vendors) that gives the users an opportunity to share knowledge they have gained in using a particular system, to exchange programs they have developed, and to jointly influence vendor software, hardware, support, and policy.

vacuum feeder   A type of feeder usually employed on floor-type offset duplicators. Vacuum tubes or "suction feet" are used to pick up each individual sheet of paper from the feed table and pass it into the impression system.

vacuum frame   An exposure table employed in making offset plates. During exposure, a film negative or positive is held in contact with plate material by a vacuum.

value analysis   A systematic analysis of each component of a product and of the operations performed on each to determine whether the value contributed by each is great enough to justify the cost.

variable   A segment of a prerecorded document that is subject to change. Examples are the name and address listings merged with form letters, or the specific information merged to produce a complete will or other legal document.

variable expense   The portion of expense that should vary in proportion to and because of a change in the volume of activity, as indicated by a determinant such as sales or production. This characteristic is determined by a study of the behavior characteristics of the expense.

variable text   Text of a changing nature (keyboarded or prerecorded), which may be combined with recorded text such as selected paragraphs to form a complete document.

variance   The difference between a budget allowance, or a standard cost, and the quantity representing actual performance.

varityper   A direct impression composition device. Type sizes generally range from 6 to 12 points.

VDT   Video Display Terminal. A CRT or gas plasma tube display screen terminal or keyboard console that allows keyed or stored text to be viewed for manipulation or editing.

velocity   Speed and force at which a typing element strikes the platen.

vendor   A company that supplies office equipment, software, or supplies.

vertical files   Vertically stacked file drawers in which files are arranged from front to rear.

vertical scrolling   The ability to move vertically, a line at a time, up and down through a display page or more of text. Allows text which will not fit on a video display screen to be accessed for review or editing. Many systems have a display buffer (memory area) larger than the display screen capacity (e.g., the screen might be 66 lines, while the display buffer holds and may vertically scroll through 99 lines).

voice bank   A recorder system whereby spoken material can be stored and easily accessed for reference.

voice compression   A device that compresses speech into a shorter time interval than the original recording.

voice-grade channel   Typically a telephone circuit normally used for speech communication, and accommodating frequencies from 300 to 3,000 Hz. Up to 10,000 Hz can be transmitted.

volatile   Refers to a type of memory which, if power has been disconnected, does not retain information.

VOR   Voice-Operated Relay. A device that activates the recording mechanism of a dictation machine by a voice sound, and by which the absence of speech over the line will cause the unit to cease recording until speech resumes.

VSC   Variable Speed Control. Used in voice compression machines.

wage and salary administration   Maintaining a logical salary and/or wage structure.

wage and salary survey   A survey of what other companies are paying for various jobs.

WATS   Wide Area Telephone Service. A service provided by telephone companies which permits a customer, by use of an access line, to make calls to telephones in a specific zone on a dial basis for a flat monthly charge. Monthly charges are based on the size of the area in which the calls are placed, not on the number or length of calls. Under the WATS arrangement, the U.S. is divided into six zones to be called on a full-time or measured-time basis.

web   Roll paper supply for floor-type offset duplicators.

weight (paper)   See *basis weight.*

weighted average   An averaging technique where the data to be averaged are multiplied by different factors. For example, a regular average is equivalent to a 50–50 weighted average. An average could be made up by taking 90% of one figure and 10% of another figure. This would then be a weighted average. Note that the weights must always total 100%.

whiteprint   A print made on a diazo duplicator. The term was coined in reference to blueprint, an older technique.

whiteprinter   A diazo duplicator. See *diazo process.*

wideband   A communications channel having a bandwidth characterized by data transmission speeds of 10,000 to over 1 million bits per second.

widow/orphan   See *automatic widow adjust.*

Winchester disk   See *magnetic media.*

window envelopes   Envelopes designed so that the inside address of the letter appears through the window so that the envelope does not need to be addressed.

wipe-on plate   An offset plate that is coated with a light-sensitive material, usually diazo sensitizer, just prior to use. The plate type is aluminum. Contact/exposure is required, followed by a one-step development process to bring out the image. See *diazo process; plate.*

word originator   See *principal.*

word processing   The transformation of ideas and information into a readable form through the management of personnel, procedures, and equipment to provide faster and more efficient business operations.

word processing center   The room or area with equipment and personnel for systematically processing written communications.

word processing system   Refers to the specific hardware, software, and peripheral devices employed to perform word processing tasks.

word processing systems manager   The individual who has responsibility for conducting a word processing study and implementing and managing a word processing system.

word processor   A typewriter that records typing onto a medium or into memory for later playback.

words   Commercial schools usually express typing speeds in "words," a word being 5 strokes (periods and spaces), so that a person typing 40 words per minute is considered to have struck 200 characters in one minute. It is not considered a suitable measurement.

work count   A count of an employee's work volume.

work distribution chart   A consolidation of task lists and activity lists to show what a department does and how each worker fits into the department's activities.

work measurement   A process of determining how much time is required to do a given amount of work.

work sampling   A work measurement technique that uses random sampling to determine the amount of time spent performing various activities.

work simplification   A planned approach to simplifying clerical work and thus increasing production per man-hour.

work standard   The time required to complete a task, as determined by a work measurement study.

workstation   A basic physical unit of a word processing system which may be composed of such hardware as display unit, keyboard, and media drive(s), and which allows the operator to perform word processing (and perhaps other) tasks.

WP   See *word processing*.

WPM   Words Per Minute. A measure of print or transmission speed, usually computed on the basis of six characters (five plus a space) per word.

WPS   Word Processing Society. A large independent word processing association (not affiliated with IWP) headquartered in Milwaukee.

WP typewriter   See *magnetic keyboard*.

writing line   The maximum horizontal space within which a typewriter will print.

X.25   The protocol recommended by the CCITT as the standard for international transmission of data over telecommunications lines.

xerographic printer   See *electrostatic printer; electrostatic process*.

ZnO   Zinc oxide. A possible coating for a photoconductor, commonly used on coated paper for copiers. See *coated-paper process; photoconductor*.

Index